THE COMPLETE HANDBOOK
OF
BEE-KEEPING

Herbert Mace

THE COMPLETE HANDBOOK OF BEE-KEEPING

New, fully revised edition prepared by Owen Meyer,
General Secretary of the British Bee-keepers' Association

Foreword by Arthur M. Dines,
President of the British Bee-keepers' Association

 VAN NOSTRAND REINHOLD COMPANY
NEW YORK CINCINNATI TORONTO LONDON MELBOURNE

Van Nostrand Reinhold Company gratefully
acknowledges Mr. F. Q. (Quint) Bunch,
former secretary of regional and state
bee-keepers' organizations and member of the
American Bee-keeping Federation, who
read and checked the text for the American
edition and provided the list of suppliers in the
United States.

Ward Lock Limited would like to thank J.
Spiller and M. G. Apiaries for permission
to reproduce several line drawings;
McGraw-Hill Book Co. Ltd. for permission
to reproduce the anatomy drawings which are
after Snodgrass: *Anatomy and Physiology of the
Honeybee*; Mr. Graham Burtt for the loan of
blocks of appliances; Eric Greenwood and
Karl Showler for their kind help in supplying
photographs.

Library of Congress Catalog Card Number 76–15804
ISBN 0–442–25045–2 (cloth)
ISBN 0–442–25046–0 (paper)
First published in Great Britain 1976
Reprinted 1976 by
Ward Lock Limited, 116 Baker Street,
London, W1M 2BB, a member of the Pentos
Group.

Printed in Great Britain

Published in 1976 by Van Nostrand Reinhold
Company. A division of Litton Educational
Publishing, Inc. 450 West 33rd Street, New
York, NY 10001

16 15 14 13 12 11 10 9 8 7 6 5 4 3 2 1

**Library of Congress Cataloging in
Publication Data**

Mace, Herbert.
    The complete handbook of bee-keeping.

    First ed. published in 1952 under title: The
bee-keeper's handbook.
    Bibliography: p.
    Includes index.
    1. Bee culture. 2. Bee culture—Great
Britain.
    I. Meyer, Owen. II. Title.
SF525.M27 1976   638'.1   76–15804
    ISBN 0–442–25045–2
    ISBN 0–442–25046–0 pbk.

Text filmset in Baskerville
Printed Offset Litho and bound by
Cox & Wyman Ltd, London, Fakenham
and Reading

Contents

Foreword

It is indeed a pleasure for me to write a foreword to this new edition of a book that will, I am sure, meet a present demand.

I have had happy associations with the two people principally concerned. In my own early days of bee-keeping Herbert Mace gave me a lot of advice and practical help; I visited his apiary at Harlow several times. His knowledge and enthusiasm were communicated widely through his writings. This book, the original of which was published in 1952, was the culmination of a lifetime of study and experience.

The pattern of the honeybee's life was established many millions of years ago, so it is unlikely to have changed much during the past twenty-five. Neither are bee-keepers all that different; but there are factors that made some revision of the book necessary. The results of continuous scientific study of the bee in various parts of the world have, in a number of ways, changed our understanding of its complex existence. The new knowledge has led to some changes in apiary practice. There have been advances in our knowledge of bee diseases too, again leading to change in approach.

The great technological developments in our own lives, their effect on the environment and, above all, the revolution in farming practice, are bound to have impinged on bee-keeping.

The revision called for someone in touch with modern trends and familiar with the needs and problems of the average reader of such a book. This has been done under the guidance of my friend Owen Meyer, General Secretary of the British Bee-keepers' Association. Nevertheless, the major part of the text remains in Herbert Mace's own words, in his own style, and still arranged in the logical manner of the original presentation.

He wrote mainly for amateur bee-keepers, at all levels. They after all comprise the overwhelming majority in Britain, as in most other countries. As one who relied on his bees for part of his living, he was however not unaware of the commercial aspect. He does give it some consideration; but on the whole his advice is to those who keep bees to get a modest amount of honey and a lot of pleasure.

The views expressed are not dogmatic; in fact I can remember Mace as being a voluble opponent of those who were considered to be so in years gone by. Whilst stating his preferences on such things as hives and methods of management, alternatives are discussed in an unbiased manner. There will be critics who complain of old-fashioned touches that do remain in the book, but others will delight in them. Individuality is one of the characteristics of bee-keepers.

For the possible, or prospective, keeper of bees, here is a new world to be explored. Not all of it is likely to be comprehended at once—enlightenment will come with experience—and one imagines the book coming off the shelf many times.

Those who are established, with some experience, can still dip into it as problems arise or a particular phase of interest has been reached. In that sense it is a Complete Handbook of Bee-keeping.

Arthur M. Dines
Canterbury, England

Part I

1
Methods of Bee-keeping

PRIMITIVE BEE-KEEPING

Honey is the sweetest, most nutritious natural food and has always been
eagerly sought. In a Spanish cave there is a painting by a Stone Age man,
depicting men climbing up to a hole in the rock to collect honey, apparently
indifferent to the swarm of bees buzzing around them. Two important things
seem to have been discovered in very early times: namely that bees are afraid
of smoke and become harmless under its influence, and that a swarm put into a
suitable receptacle will stay there. Some of the most primitive African races
still keep bees in hollowed logs which they hang in trees.

Civilized people improved on this. In Persia bees were kept in chambers
scooped out of the thick house-walls, a small hole opening outwards for the
bees' flight, and a door inside permitting the removal of honey. The Egyptians
used hives made of clay vessels like drainpipes, piled on one another, so that a
large number could be kept in a small space. These can still be found by the
Nile in active use.

HIVES—ALL SHAPES AND SIZES

In Europe, the form of hive which seems to have been used in Greek and
Roman times was dome-shaped. At first it was of wicker, plastered with mud
and cowdung, but in the West this gave place to straw, which is cheap, light,
dry and warm. Tall domes were at first general, and are still in favour in
Holland, but in Britain the hive tended to be dwarf. The first step towards
improvement was the use of a shallow tray fitted over the top in summer, as it
was found that the bees put their purest honey at the top of the hive. This
made it necessary to have the main body of the hive flat on top, with a hole in
the centre for the bees to pass through, this being stopped with a bung when
the upper storey was not in use. At one time it was usual to put a bell glass over
the hole and one of these glasses, filled with honey in virgin comb, was a great
prize.

This simple arrangement was the prevailing method of keeping bees down
to the middle of the nineteenth century, for although new and better methods
had been invented long before this, they were very slow to reach the peasantry
who comprised the larger number of bee-keepers. Another reason for slow
progress was the fact that bees are not tameable like cows and sheep, ducks
and hens. To be healthy, they must be free to come and go, and there is nothing to prevent them leaving the hive altogether and going wild. Indeed, from
Roman times it has been the law that if a swarm goes away, and the owner
does not keep it in sight, it is no longer his property. Anyone who can capture
it can keep it.

CONTINUITY OF WORK

There is evidence that the Greeks were aware of a trait in bees which was ultimately to prove the most important factor in modern management, but it was not until the eighteenth century that its significance was recognized. This is the fact that bees always prefer to continue existing work rather than start afresh; if a piece of honeycomb is stuck in the roof of a hive, no matter of what construction, they will use it as a foundation for a full-sized comb.

Advanced bee-keepers began to use hives opening either at the top or the back, with narrow bars resting at the top. To each bar a piece of comb was fixed, so that the bees built each full comb from this foundation. They attached them to the *side* of the hive, but by cutting through these attachments, the bars, with the adhering comb, could be taken out and replaced. Wicker or straw was still used in construction, but hives were altered in shape for convenience in affixing the comb bars and ultimately all movable-comb hives became rectangular, a shape more suitable for construction with wood than with the older materials.

'BEE-SPACE'

The year 1789 marked the greatest revolution in bee-keeping methods, for it was then that François Huber, who made the most astonishing discoveries about bees, invented his famous leaf hive, which consisted of wooden frames hinged together like the leaves of a book, so that they could be separated and examined on both sides, without serious disturbance of the bees.

During the next half-century, many fanciful hives, based on Huber's discovery, were tried, but it was not until 1851 that the open-topped hive of the American, Langstroth, was devised. After many experiments with hives containing frames of the old type, which had to be cut out each time they were removed, he discovered that if the space between the outside of the frame and the inside of the hive is about one quarter of an inch, the bees do not fill it in, but use it as a passage-way. Any larger space they fill with comb, while a smaller one is blocked up with 'propolis' or bee gum.

From this time onward, all beehives, no matter how they varied in other respects, conformed to this principle. The frames were made with projecting ends at the top, which rested on rebates in the wall of the hive and, being half an inch smaller from side to side than the inside of the hive, hung so that the 'bee-space' as it was now called, kept them all free from side attachments: they could be taken out as required. Designs and styles have become very numerous, but all experienced men agree that the simpler the design the better. Those favoured at present are described in the chapter on hives.

The earlier frame hives were provided with a loose board, known as the Crown board, as a covering for the frames. This was sometimes so firmly stuck down by the bees that it was difficult to remove without damaging the combs. After a time, this board was displaced by a 'quilt' of American cloth or calico. This was less trouble because it could be turned back and stripped from each frame successively, but the bees dislike anything loose and yielding and try to remove it, so that the cloth is soon bitten into holes. They also add much gum to it and fasten it down to the frames, which makes it difficult to remove. Quilts also blow about when removed in windy weather and, worst of all, they prevent bees passing over the frames, so that special 'winter passages' have to be put under them in the autumn. Although many bee-keepers still use quilts, an increasing number have returned to an improved form of Crown board. In the United States the quilt has been replaced by an inner cover, with bee space provided to allow free passage over the tops of the frames.

In some countries, notably Switzerland, Germany and Austria, hives are kept in houses or sheds containing any number. These have several advantages. They enable work to be done in wet or windy weather, ensure perfect dryness at all times, and are actually more economical to construct.

COMB-FOUNDATION

Next to the movable-comb hive, the most useful invention was wax foundation. A few years after Langstroth's frame hive was produced, J. Mehring of Germany made a wax sheet, stamped with a rough impression of the bases of comb cells. It was found that bees used this as readily as natural comb and gradual improvement of these sheets has added greatly to efficient bee-keeping. Not only does foundation enable combs to be built perfectly flat and level, but the number of drone cells is much reduced. Frames are now made with great accuracy by machinery and have reasonable strength combined with light- ness. Several attempts have been made to introduce foundation, or even com- plete combs, made of aluminium, but they are very costly and do not seem to have any advantage over natural comb.

The last important invention I shall mention here is the honey extractor. The original inventor was an Austrian, de Hruschka, who in 1865 devised a machine which threw honey out of the combs when they were revolved rapidly. This enabled combs to be used again and again, saving time, wax and bee labour. Several types are now in use and will be described later.

MODERN METHODS

Before the sugar cane began to be widely used in the production of a sweetening agent, honey was practically the only food of its kind in general use. Its virtues are frequently extolled in the Bible, and Samson's riddle 'Out of the strong came forth sweetness' is well known. In classical and medieval times beehives were a regular feature of country houses and farms, and they were considered of sufficient importance to be enumerated in the Domesday Book, along with other stock.

Their management was simple, for few knew, or desired to know, anything about the interior of the hive. In spring or early summer, a watch was kept for swarms and, when they issued, they were shaken into an empty hive or 'skep', put on a stand or 'stool' with the other stocks, and all were left severely alone till autumn. They were then sorted over and those heavy with honey were 'taken up'. A hole was dug, burning sulphur placed in it and the skep stood over it till the bees were dead. The comb was cut out and strained through muslin. The mass of wax was then put into a vessel of tepid water and soaked till the remaining honey was loosened and washed out. The liquor was used to make mead and the wax dried and melted down. Though the quantity of wax was relatively small, it was more valuable than the honey, for in those days the only form of artificial light came from candles.

Though this simple form of bee-keeping has often been cited as cruel and barbarous, when carried out according to traditional rules it was quite sound, because the selection of the heaviest hives for 'taking up', meant that the swarms—almost invariably the ones gathering the largest amount of honey—were disposed of, and as these contained old bees which would not, in any case, survive the winter, it was not unmerciful to kill them before it came. Stocks of medium weight had young queens and an adequate store of honey to last through the winter. Very light stocks without sufficient store were sometimes taken up with the heavy ones.

Straw skep hives were widely used before frame hives

DRIVING

When the new frame hives had become established as the approved method of housing bees, someone conceived the bright idea of getting bees for nothing by 'driving' bees from cottagers' straw skeps. If a skep is given a good dose of smoke and turned upside down, the bees can be persuaded, by drumming steadily on its side, to leave it in a body. By arranging an empty skep at a suitable angle above the full one, the bees can be collected in it, leaving their brood and honey behind. This saved the cottager the trouble and pain of killing the bees and left the skep free from dead ones, so it became the regular practice for 'experts' to tour the countryside in autumn, driving cottagers' bees and either feeding them on sugar syrup, or selling them for other bee-keepers to do the same. For a long time this was a lucrative business, especially in the south of England, where skep bee-keeping was carried on freely long after frame hives became general. An end was abruptly put to it by the outbreak of 'Isle of Wight disease', because these old bees were the most liable to have the disease and they did, in fact, help very much to spread it to all parts of Britain. When this was realized, the market for driven bees slumped, and since most of the old skeppists lost their bees from the same cause, this old method was brought to an end. Just here and there you may still find an old man who keeps bees in this way, selling his swarms instead of keeping them for honey production. Bees have become so much more valuable since the war that it pays him better and is less trouble.

This inverted skep shows how bees build and attach their combs. The large cells at the lower edge on one comb are queen cells

HELP FROM THE BEES

The introduction of sugar from the cane and later from beet, ultimately reduced honey from its old status as a necessity to a luxury, and by the nineteenth century bee-keeping declined rapidly from being a normal part of agriculture to a hobby. In this direction it had become more appealing because of the discoveries of Huber and his contemporaries, and for a long time it was carried on by country parsons and others interested in natural science. This tendency was fostered by the high cost of the 'newfangled' hives. Any intelligent rural craftsman could make a skep from a bundle of straw and a few bramble stems.

The formation of the British Bee-keepers' Association, and later of county and district associations, led to wider interest and bees began to be kept, not only in the country, but also in suburban gardens and even large towns. About the same time, increasing knowledge of food values brought to light the fact that honey is far more valuable as a foodstuff than commercial sugar, and the demand began to grow. Imports of honey from America, where progress in bee-keeping methods has been more rapid and crops larger than in Britain advertised the virtues of honey, and thus the possibility of honey production as a means of livelihood began to be recognized.

Perhaps the most important reason for the increased interest now taken in bee-keeping was the discovery that the value of bees as producers of honey and wax is small compared with their rôle as pollinators of flowers. It has now been fully demonstrated that apples, pears, plums and indeed all hardy fruits give much heavier crops when bees are kept in the orchards. Not only do many fruit growers keep bees solely for this purpose, but those who do not care to manage them often hire stocks from bee-keepers, who put them in the orchards during blossom time. What is true of fruit is equally true of any crops, particularly the clovers, grown for seed. With bees on the spot, incomparably larger quantities of seed are produced.

Over the course of evolution bees and flowering plants have developed mutually advantageous features. The bee's food consists entirely of pollen and nectar so she is compelled to visit open blossom for her sustenance. The average colony of honeybees will consume between 50 and 100 lb of pollen (their source of protein and vitamins) during the year. The honeybee is physically well adapted for the collection of pollen, since she is covered with plumose body hairs to which the pollen grains adhere. A pollen gatherer will move quickly among the stamens biting at them with her mandibles in order to dislodge the grains and gathering them to herself with her legs. If she is collecting nectar she will push past the anthers to reach the nectaries. In either event she can hardly fail to dislodge some pollen onto the sticky surface of the stigma.

The economic importance of efficient pollination of seed crops can hardly be over-emphasized but perhaps one example will suffice, that of field beans. Recently six years' caging trials on this crop showed a mean gain of 11 cwt per acre of beans where bees had access to the crop (yield average $33\frac{1}{4}$ cwt/acre) compared with bees excluded (average $22\frac{1}{4}$ cwt/acre). Counting good and bad years, the overall gain from bees was $8\frac{1}{4}$ cwt/acre.

PROPOLIS

Yet another product of the hive which is becoming increasingly important, is the resinous substance exuded from the buds and leaf scales of certain plants and trees and collected by bees for filling unwanted crevices in the hive. This is known as 'propolis' (literally 'before the city'—Greek). The name is said to derive from its use by certain races of bees, notably Caucasian, in the building of walls just inside the entrance.

Its use as an ingredient in the varnish used on Stradivari and Amati violins is well known, but since ancient times propolis has been used in folk medicine for its alleged therapeutic properties. It was administered in the Boer War in wound treatment and has been widely used medicinally in Eastern Europe where it is said to have beneficial effects in cases of throat infection, catarrh, infectious wounds, bladder and kidney complaints and for the promotion of general well-being. There seems to be some evidence that it contains an antibiotic substance, not yet isolated. Propolis is at present sold by the ounce at very attractive prices based on a grading system.

QUALITIES OF A BEE-KEEPER

Bee-keeping by modern methods undoubtedly demands more personal interest. Though still far from requiring the attention demanded by other forms of stock, which usually have to be fed and watered at least once a day, any necessary assistance must be promptly and efficiently given. To be a successful bee-keeper requires courage, patience and gentleness. With these qualities anyone may keep bees with reasonable prospect of success, even in some of our large towns. There are apiaries on rooftops in the vicinity of the London parks, where quite a fair quantity of surplus honey is taken every year, though it is naturally not equal to crops secured in favoured places such as the heather moors of England and Scotland.

Even if no pecuniary reward resulted, the enthusiasm seen among bee-keepers at honey shows or field meetings indicates that as a hobby of surpassing interest bee-keeping has no equal.

2
The Bee's Place in Nature

Bee-keeping *can* be started without any capital worth mentioning; for the most ignorant and poor person by catching a stray swarm and putting it in a box, may become a kind of bee-keeper. Quite a few have, in fact, started like that and, becoming interested, have gone on to make a success of the craft by learning about new methods and studying the habits of their charges. The beginner should make himself acquainted as soon as possible with the habits of bees, so that he may know when to feed them, when to enlarge their accommodation, or when swarms may be expected. Otherwise he will soon be involved in unprofitable labour and even loss.

The honeybee belongs to the insect order *Hymenoptera*, which contains many thousands of species of bees, wasps, ants, and some allied species which do not interest us at the moment. Their principal characteristic is the possession of two pairs of membranous wings, the front pair much larger than the hind pair. These hind wings have, along their front edge, a row of minute hooks, which can be fixed into a groove on the hind margin of the front wings. This provides a broad powerful surface in flight, but the hooks are instantly detachable and the separated wings can be folded close to the insect's body. Since many of the members of the order rear their young in narrow cells, this arrangement is very convenient, if not essential. Another prominent feature in this group is the possession of an abdominal sting, with which many species can inject a powerful poison into the bodies of their enemies or victims. This weapon is most frequent in the wasps and bees, which have other features in common. Both have legless larvæ, which are reared in specially constructed cells. All the wasps, both solitary and social species, make the cells of alien materials, earth, stones and the like made into a kind of cement with saliva, or of wood fibre pulped and fashioned into a delicate paper-like fabric. The solitary bees use similar materials, several cutting out pieces of leaf, or boring into wood.

WASPS AND BEES

The main distinction between wasps and bees is that the food of wasps is nectar or fruit juice and animal food, such as flies, caterpillars or even defunct birds and mammals, while bees of all species live solely on the product of flowers— nectar and pollen. To correspond with these different habits, their respective structures are modified. Wasps have sharper jaws, almost naked bodies and are more agile than bees, which are more or less hairy.

SOLITARY BEES

Bees are roughly divided into two groups, solitary and social species. The solitary bees consist of males and females. After pairing, the females construct cells in a large variety of ways. Some bore a tunnel in the earth (mining bees), others choose the hollow stems of plants, or bore into wood (carpenter bees). The cells are made in these tunnels by lining the walls with such material as pieces of leaf or grains of earth or sawdust cemented together with saliva. A mass of honey and pollen is placed at the bottom of the cell, an egg laid in it and a partition of the same material as the wall-lining closes it in. Another similar cell is provisioned and provided with an egg and so on till the tunnel is filled, when the bee's task is finished: she has no further concern with her offspring, who in due course grow to maturity and ultimately emerge as male or female bees.

BUMBLE BEES

Social bees live in colonies or families containing, in addition to the males and females, a third form of adult known as 'worker', really a female whose reproductive organs are undeveloped. The typical species of social bees are those in the genera *Bombus* (bumble bee) and *Apis* (honeybee). The bumble bee colony is an annual one, started in spring by a fertile female, who makes a nest of grass or moss, sometimes on the surface, but frequently in a hole in the ground. In this nest she makes several cells or waxen pots, in which she stores honey and pollen, and other cells in each of which she lays an egg. She sits upon these cells until the larvæ are hatched, and then feeds them till they have completed their growth and sealed themselves in the cell with a silken cover. In due course, worker bees come from these cells and take over the work of collecting honey and pollen, leaving the female or queen bee to lay eggs and attend the young. With this force of workers, the nest grows rapidly, but none of the bumble bees produce more than 80 to 120 workers. In autumn, male and female bees are produced. These pair off and the females hide themselves in some snug spot for the winter, beginning the work all over again the following spring.

Since the colony is annual only, bumble bees do not store more food than is needed for use from day to day and it is only in the genus *Apis* that we find colonies which are perennial and able to tide over periods when no flowers are in bloom by collecting more food than they require for immediate use. Several species and sub-species of *Apis* are known, but the only one of paramount interest to humanity is *Apis mellifera* Linn., an insect whose way of life is so complex and highly organized that it has excited the wonder of mankind throughout history.

THE BEE COMMUNITY

A swarm of bees, such as most country people have seen hanging on a bough, is a roughly pear-shaped cluster of brown insects, clinging closely together, a feat they perform by attaching to each other and the object they are resting on, the hooks with which their feet are provided. A swarm may contain any number of individuals, from five- to thirty-thousand, but strictly speaking, it is only one creature, in spite of the fact that the parts are able to separate from the mass and return. The bee colony is, indeed, much like a tree. A tree is made up of roots, stem, leaves and flowers, each of which plays a part, the principal one being that performed by the leaves, which alone can collect food from the air and turn it into plant tissue. In the same way, worker bees collect food and elaborate it for the nourishment of the colony.

In a bee colony there is only one female or queen, who may be compared with the growing point of the tree and, in both cases, the rate of growth is conditioned by the amount of food brought in, by the workers in one case and the leaves in the other. The flowers of a tree may be likened to the drones or male bees, which are usually produced only at a certain season of the year and are, like flowers, the first steps towards reproduction. The fruit of a tree is paralleled in the bee colony by the young queens or females which are produced at the peak of the season. Each and all these members of a bee colony are dependent on the others and cannot exist separately.

Worker

Drone

Queen

WORKERS

The workers are the most numerous and familiar to everyone. They are noteriously busy all the sunny days of summer, collecting honey and pollen, and storing it in waxen cells, which they construct from a kind of fat secreted in glands beneath the body and not, as used to be thought, collected from flowers. In the same cells, they rear the young, feeding them diligently, at first with a glandular secretion or 'milk' and later with a mixture of honey and pollen. Water is also collected from moist places and propolis or bee gum is scraped from the buds of trees for use in reinforcing the comb and filling up cracks in the walls of the hive. Workers are swift and direct in flight and can carry surprisingly heavy loads.

To perform these and other tasks they have a remarkable collection of natural tools. The head is provided with strong jaws, a pair of large compound eyes, giving a wide range of vision, sensitive antennæ, which are the organs of touch and probably other senses not fully understood, and above all, one of the most remarkable tongues found in nature, enabling the insect to collect liquid of varying density. It can act either as a spoon for licking up small drops, or a pump to draw in large quantities rapidly. This tongue is the gateway to the alimentary canal, leading to a dilatable bag, which corresponds to the paunch of ruminating animals: the honey stored in it can either be brought up again and deposited in the comb-cells, or passed on to the true digestive stomach.

The thorax, or chest, carries the two pairs of membranous wings, which can be connected in flight, or disengaged to lie flat against the body when the cells are entered. There are three pairs of legs, each bearing hooked feet, by which the creatures can attach themselves to each other, or cling to almost any surface. The legs have combs and brushes for removing pollen from flowers or the hairs of the body. Most remarkable of all is the *corbicula*, or pollen basket of the hind leg, in which the bee packs pollen and carries it home.

On the underside of the abdomen are the wax pockets from which wax exudes in small scales, which are removed by the legs and used in comb building. At the tip of the abdomen is the sting, an extremely fine and tough piercing instrument, operated by a set of powerful muscles and connected with a bag containing the poison which produces such painful effects in any unfortunate victim.

Not least remarkable are the numerous glands in various parts of the body, each producing its own special secretion: from wax for comb building to the 'royal jelly' used for raising queen bees.

The life-span of a worker bee depends on the time of year. Those hatched in spring or summer are rapidly worn out by ceaseless activity and on the average do not live more than six weeks, but workers raised in autumn live through the winter in a semi-dormant state and are able to carry on the work of the hive in the early days of spring.

DRONES

The drones or male bees are easily distinguished by their somewhat greater size, stout hairy bodies and large eyes. They are not present at all times, but appear first about April and usually number only a few hundreds. They have none of the elaborate organs of the workers and do not gather honey or pollen, but live on the food stored by the workers and remain in the hive, except during the best part of sunny days. They then take short orientation flights, flying round with a deep boom, quite different from the hum of busy workers. Their main purpose is to fertilize the queens. When a virgin queen leaves her hive on the 'wedding flight' she is pursued by the drones in her neighbourhood. Drones are capable of powerful flight and it is believed that they must frequently range miles from home, for queens are often mated in places where no drones are known to be. After mating, the drone dies, his genitals being

Drone comb and eggs. There are four of these cells to an inch under normal conditions

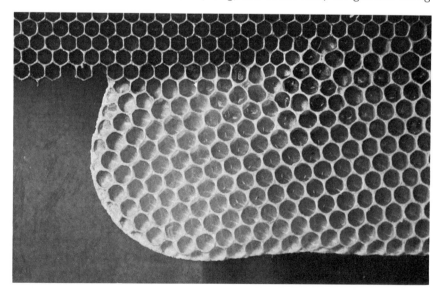

ruptured and retained by the queen. Drones have no sting and cannot defend themselves against the workers, who ruthlessly cast them out of the hive when there is a dearth of food supplies. Only under abnormal conditions, such as the absence of a queen, are they allowed to live through the winter.

THE QUEEN

However large a swarm may be, it is quite useless unless it has a fertile queen. In ancient times, before her true nature was ascertained, this remarkable insect was called the 'king' bee and was supposed to be, like a human despot, the ruler of the hive. Indirectly, this is true, since no bee colony can thrive unless the queen fulfils her function of egg-laying, but it is now certain that this is her sole function, and that hive policy, if we may so describe it, is directed by what can best be called a communal spirit, actuated by surrounding conditions.

Though a queen bee is the largest occupant of the hive, her extra bulk consists solely in the greatly enlarged abdomen, which holds a pair of ovaries containing thousands of eggs in various stages of development. She has none of the apparatus for building comb, or for collecting honey and pollen. It has been established that multiple mating is normal and she may mate with up to nine drones, either on her first or subsequent mating flights. The sperm received from a single mating is rarely sufficient to fill her spermatheca. The male sperm is kept alive in her abdomen and used to fertilize the eggs as they mature. In the natural way, a queen may live five or more years, but after the second or third season her egg-laying diminishes rapidly; before this stage is reached, the workers often depose her and raise a new queen.

This brief sketch of the adult occupants of the hive is sufficient to enable the beginner to understand the kind of creature with which he is dealing, but later in the book I shall have more to say about the anatomy and physiology of the bee.

RACES OF BEES

The honeybee is a native of the Old World and was not known in the New till introduced by European settlers. The exact country of origin has never been determined, but a few fossil bees, much like *Apis mellifera*, have been found, mostly in Germany. There is no doubt that the species was widely distributed in Europe, Asia and Africa long before mankind appeared and, as in other animals and plants, its appearance and habits have been modified to suit each environment, so that to-day we recognize two distinct types, respectively known as black and yellow bees. Broadly speaking, black varieties are found in the cooler northern regions and yellow ones in the warmer parts.

Each of these two types has been further modified by isolation in different districts, and varieties with well-defined characters are known by names corresponding with their native places. Many of these varieties have, at one time or another, been introduced into Britain and, because the difficulties of keeping them pure are almost insuperable, there has been such complex crossing that it is impossible to relate the bees now common in these islands to any of the named varieties. It is easy enough to procure queens of a desired variety from their native home but, unless put in some spot like an island many miles from the mainland where there are no other bees, it is certain that, within a few seasons, they will become mongrels. At present there is no obvious solution of this problem and those who wish to keep a pure race must continually import new queens. It may be that the technique of artificial insemination will in time be simplified so that it may be possible to control mating in the ordinary apiary. At present, it is only practicable for the specialist, who can work in controlled conditions.

Before this influx of aliens, which began with the importation of Ligurian bees into Britain in 1859, our bees were a well-marked black variety. Their appearance was delineated in a famous plate in *The Management of Bees* by Samuel Bagster, Jr., published in 1834, and old hands at bee-keeping remember them quite well. Although it is extremely doubtful whether any have survived the various epidemics and importations, the type still remains the standard of comparison by which we judge the merits of any stock of bees. I will therefore begin with a description of these departed and, by some, lamented bees.

ENGLISH BLACKS

Though not really black, they were very dark brown in colour, active on the wing, hardy and long lived. They wintered well on very moderate stores and were almost indifferent to weather conditions in summer, working in cool, cloudy weather as well as in sunshine and warmth. They made comb of great beauty, with white even cappings, and so were esteemed for the production of sections. The queens were not heavy breeders and in times of dearth breeding decreased or stopped at once. They swarmed only moderately. They were little disposed to sting if suitably handled, and one of their faults was that they did not defend the hive too well against robbers. Although not vicious, they were restless and excitable under manipulation, flying from the combs in panic and creating an uproar. They were not very good house cleaners and were, perhaps because of that, susceptible to disease. Indeed, it was the readiness with which they went down under 'Isle of Wight disease' in the first years of this century that finished them. Not only were they wiped out in many parts of the country, but the importation of Dutch and other races to make good the losses, mongrelized the rest. However, it must be said that a number of competent bee-keepers have claimed the survival of the old native blacks in isolated wild colonies.

DUTCH BEES

In 1914 Mr J. C. Bee-Mason imported into Britain a large number of stocks from Holland, since he had tried and found them resistant to the prevailing epidemic. Through Mr Mason's kindness, I had one myself and I found the bees active, vigorous workers during the first season. In the following year they began to swarm incontinently. It seemed impossible to stop them during the whole of the swarming season and, in the moderate honey-flow of that year, there was not a stock strong enough to get surplus. I went abroad the next year, but my wife had the same trouble with swarming, although the crop was fairly good. She disposed of most of the stocks, leaving three in charge of a neighbour, who reported at the end of the war that all had died of the disease.

Everyone who has tried them confirms my experience. They are hardy and prolific, but this persistent swarming renders them useless.

FRENCH BLACKS

Some time after the first world war, there was a vogue in Britain for 'package bees' sent from France. These were 'driven' bees such as were, before the epidemic, regularly sent around the country from skep apiaries. These bees came mainly from the Le Gatinais district and many hundreds of packages were distributed. There is a wide difference of opinion about them. I had some and they made no particular impression on me. They were just ordinary bees, a little more prolific than the old blacks and with the same habit of leaving the combs during manipulations. I do not recall that they were specially ill-tempered, but most apiarists, including Mr R. O. B. Manley, one of our largest commercial bee-keepers, complained that they were so vicious as to be useless for his work. Mr William Hamilton, on the other hand, in the *Art of Beekeeping*, praises them as the best bees of all—if their bad temper could be eradicated.

CARNIOLANS

These were at one time extremely popular, because of their exceptional gentleness. Their home is Austria and Jugoslavia and they are distinguished by rings of white down on the abdomen. They are rather large bees, exceptionally easy to handle, because they stay quietly on the combs. They have a good reputation as honey gatherers and are very prolific. They come next to Dutch in disposition to swarm, which is probably why they have not been widely adopted.

CAUCASIANS

These are comparative newcomers to Britain, a large number having been imported during the 1930s. They are somewhat like Carniolans in appearance, black with whitish hairs. I had three queens which built up strong colonies quickly and I found them hard workers even during poor weather. They capped their comb nicely, were quite good tempered and wintered well. They did not swarm excessively, but when they did I was disconcerted to find that they went to the tops of the tallest trees in the vicinity. I saw no reason to prefer them to common mongrels and in time they faded out as pure-breds. I did not notice what many complain of—the use of excessive quantities of propolis—but, surrounded as I am by trees, I find any bees use a good deal of it. Mr Manley speaks highly of them, finding them good collectors, ceasing to breed early and wintering well on moderate stores. Though gentle as a pure variety, he found that their crosses were intolerably vicious and gave them up for that reason. Mr Hamilton says they are docile on occasion, but sometimes so bad that they cannot be subdued by smoke. He did not find them prone to swarm, but as honey getters they were not satisfactory, except at the heather.

YELLOW BEES

Of the so-called yellow bees, the best known is the Italian Bee or, as it used to be called, the Ligurian. It was first imported in 1859 and since that time there has been regular importation into Britain except during the last war. Commercial bee-keepers are almost universally in favour of this variety in Britain as well as in the USA and other parts of the world. Beyond doubt, its handsome appearance is one of the chief reasons. When pure, the workers have three golden bands on the abdomen and most of the body hair is yellowish. Some have as many as five yellow bands and are not seldom mistaken for wasps by the uninitiated. The queens are also very distinctive, varying from bright yellow to leather coloured. They are the easiest queens to find, a feature much in their favour, as dark queens often cause much time waste in looking for them. They are the most prolific of all varieties and are consequently prone to swarm, though never to the disastrous extent of Dutch and Carniolans.

Their chief fault is that they continue to breed throughout the season, whether nectar is coming in or not. The result is that in poor seasons of broken weather, they use up their surplus during a period of dearth. Many a beekeeper who has had a full super of honey in early June has found, after a spell of bad weather, that it has all vanished. From this it follows that they require more stores for winter and whereas 25-30 lb was ample store for the blacks, many operators consider that Italians should have not less than forty. The individual bees are short-lived. They stay on the combs during handling and are generally good tempered, though an occasional colony is far from amiable. They need dry wintering conditions and most people consider that they need more protection in winter than blacks.

A four-way hybrid from the Italian bees, called the Starline, is obtainable from many queen breeders in the United States.

OTHER BEES

Another handsome yellow bee is the Cyprian, which some have tried to bring here, but although it is said to be a hardy forager, it is very vicious and has no special merit over Italians.

The Egyptian, Syrian, and Palestinian bees have some individual features, but nothing to commend them above Italians, and are reputed to be stingers.

As I have said, there is no practicable means of keeping any race pure in Britain, except continual importation of queens. Moreover, however good the qualities of a race may be, it does not follow that all stocks will have them in full measure.

HIVE LIFE

I mentioned earlier the swarm hanging on a bough which is the only glimpse the average person gets of the bee establishment. The questions he naturally asks are: how did it get there and what will it do?

The answer to the first is simple. It came from a hive which had reached a prosperous and overcrowded state and sent forth its surplus population to found a new home. In this, it follows the rule with all living things, except in one particular: instead of the emigrants being the young and lusty, a swarm

24

consists of old workers and the old queen, only a comparatively small number of young workers being moved to go with the expedition. From May to July is the normal time of swarming, though in exceptional circumstances it may take place earlier or later. The time of day is pretty constant, rarely before eleven in the morning, or after four in the afternoon. The exodus is usually sudden. Without warning, bees come tumbling out and fly round till the air is filled with a merry whirling cloud. After a few moments of wild excitement, the throng converges on some spot—the bough of a tree, a bush, hedge or fence; sometimes, if there are no upstanding objects around, it may settle on the ground. It quickly compacts itself into a tight mass and remains still for a varying period.

What it does next depends on circumstances. If there has been fine weather for some days preceding, it may move off in a matter of minutes and go direct to its chosen destination, for during fine weather the new home is usually found before the swarm leaves. At this time of the year one often sees bees prowling in unusual places. They investigate empty hives, holes in trees and walls and similar dark places, which they never visit in their normal search for nectar. Indeed, one of the best ways to escape from angry bees is to run into a dark shed, for they will be most unlikely to follow one there.

HOUSE-HUNTING

If the weather has been bad, the scouts may not have found a suitable place and the swarm may be forced by a sudden burst of warmth and sun to take a chance. Scouts will then leave the swarm where it hangs and search for a suitable lodging. They may be hours or even days on the hunt while the swarm hangs patiently waiting for news. If the weather turns cold and wet, the stores with which the swarm always provides itself before it sets off will be used up and in extreme cases it may starve. If the weather remains fine and a suitable home is not found, or the queen is, for some reason, unable or unwilling to move, the swarm will begin to construct comb, collect nectar and establish itself in the open. This is not very common, but I have known a full set of combs to be built on a bough, and on rare occasions the swarm has survived a mild winter.

In the usual way, a home will sooner or later be found. The watcher will observe a slight sizzling among the bees and very swiftly the swarm will take to the air and move off at a great speed, in a 'bee line' to the selected place. Hollow trees, caves, cavity walls, false roofs, chimneys, church towers and similar places are all liable to occupation by bees and some of the weirdest and, from the human standpoint, most inconvenient places have been chosen. Swarms have been found in a clock case, a piano, a pump and even a hollow metal figure, which the bees entered through a flaw in the casting. More than once has a postman had a rude shock at finding a mail-box full of bees instead of letters.

FURNISHING THE HOME

All doubt having ended, the swarm sets to work in its new home with great energy. Most of the bees ascend to the top and suspend themselves, while wax scales form in their wax glands. A few bees occupy themselves by cleaning down the walls of the dwelling; others collect gum from trees to fill up cracks in the same. These and some others, whose journey rations are used up, make themselves acquainted with the position of their new home, so that they may make no mistake and get lost. This they do by flying round in gradually increasing circles, facing the entrance at first, until every detail of its appearance is imprinted on their brains. In a surprisingly brief time they will come and go with such precision that they drop within an inch or so of the entrance every time.

These foragers return laden and hang themselves up with the cluster to make wax. As the wax scales exude, bees remove them, chew them into a plastic state and stick them to the roof, or to wax already in place. As these lumps of wax grow, they are pared away by other bees till the least more work would break through them: by this process of piling on and paring off, the comb begins to take shape. Honeycomb is one of the most remarkable features of bee life. Constructed entirely from the wax elaborated in the bee's own body, it is a marvellous example of economy in material and suitability to purpose. In it, not only is honey stored, but the young are reared and all needful work carried out in security. Yet this structure is so delicate that the walls average only ·0025 of an inch thick and a piece weighing an ounce will support as much as 3 lb of honey. Honeycomb is composed of hexagonal cells placed side by side without any space between, so that one wall serves for two cells.

Because the hanging cluster of bees acts as a plumb-line, the comb is quite vertical. It is two-sided, the central partition or 'midrib' forming the floor of opposite cells. When built for worker brood, the comb is about $\frac{7}{8}$ of an inch thick. Passage-way between combs is left, so that from centre to centre the distance is slightly under $1\frac{1}{2}$ inches. In times of heavy nectar-income cells are made deeper, often as much as 2 inches.

The openings of cells are thus on the sides of the comb and the cells slope slightly downward to the base, so that liquid placed in them cannot run out. Cells intended for raising worker bees are five to the linear inch, those for drones four to the inch. Honey may be stored in either, but during a heavy honey-flow the larger cells are made. These are naturally more economical of wax and time. When the first central comb is well under way, others are begun on either side of it and so on until the whole available space is filled.

THE NEW GENERATION

Egg-laying soon begins in the central comb, the queen putting each egg in a separate cell and proceeding outwards in a spiral, laying her eggs evenly on each side of the comb. The egg is an elongated oval, wider above, white and slightly curved. The lower end is stuck to the cell bottom. It takes three days to hatch, during which it gradually falls over till it is flat on the cell floor. On the fourth day the newly hatched larva lies curved and almost transparent in a little lake of milky fluid, the glandular food given it by the workers. During five days, it is plied with food and grows until it is a tight roll, completely filling the cell. In these five days it increases in weight 1,600 times. On the ninth day the cell is sealed over with a porous covering of wax and pollen and, inside, the larva changes from its curled-up position until it lies full length in the cell and spins itself a light shroud of silk. During the next thirteen days it shows no outward sign of life, but in the solitude of its cell changes from a legless grub to a perfect bee, coming out as a downy, winged, but rather feeble little creature on the twenty-first day from the laying of the egg.

A newly established swarm seldom raises other than worker bees. Indeed, its breeding is not very intense, since most of its energy is devoted to collecting the nectar which at this time is usually to be had in abundance. Sometimes brood is almost absent, the cells being filled so rapidly with honey that the queen has nowhere to lay. Later on, when income diminishes and the central cells are emptied, the queen begins to lay more freely, but swarms sometimes raise insufficient bees to make them safe in spring and are liable to die out for that reason.

Drones, workers or queens can be produced, the sex being decided by the queen to meet the colony's need. If, when the egg is laid, the male sperm stored in her organs is allowed to reach it, the sex is female, but eggs laid without the sperm produce males. The bees are aware of the condition of each, for drone grubs are fed differently from females. The latter are not given undigested pollen but, after the third day, drones receive a fair proportion of it. They grow much larger and need more room when changing to the pupa, so that when sealed over, the cap is domed and sticks out conspicuously. Drones are fed for six instead of five days and take two days extra to change to the adult—twenty-four days in all from the egg-laying.

Workers are reared at nearly all times of the year. If the weather is mild, breeding begins in January and continues variably till April, when it speeds up. During May and June as many as 2,000-3,000 eggs are laid daily. The brood-nest expands in a spiral for about three weeks, when the centre cells begin to empty and the queen starts there again. Drones are seldom reared before April, and not after June in an ordinary hive, but old or sterile queens produce only drones and it is a bad sign if drone brood is present at any other time than early summer.

The usual time for queens to be reared is April to July. As soon as the hive becomes crowded, by which time drones are mature, queen cells may be built. They are somewhat like an acorn in shape and size and hang down, so that instead of being reared in a horizontal cell, the queen larva hangs upside down. These cells may be built on any part of the comb and vary in number from a solitary one to more than a hundred. Perhaps a round dozen is about the average. These large cells enable the embryo queen to be fed with royal lavishness. Instead of doling out a little each day, the bees pour the royal jelly in freely. The result is that the egg which, with ordinary feeding would produce a worker, not only produces a female with fully developed organs of generation, but does so in less time; the young queen is ready to leave her cell sixteen days after the egg was laid.

Breeding Table

	Incubation	Feeding	Change to adult	Total
	Days	Days	Days	Days
Worker	3	5	13	21
Drone	3	6	15	24
Queen	3	5	7	15–16

Once the swarm has settled in its new home and stored its combs with all the honey and pollen it can gather, its activity diminishes. As autumn comes on many bees stand guard at the entrance to keep out marauding wasps and robber bees. Some continue collecting whatever is to be had in the autumn flowers, while propolis is collected freely to seal up draughty spots. Gradually the sentries grow fewer until days may pass without a bee being visible. Nevertheless, honeybees do not, like bumble bees and wasps, hibernate in complete unconsciousness. There is always some slight movement within the cluster, which expands and contracts according to the surrounding temperature. Food is taken from the combs from time to time and on any day when the temperature rises above 50 °F. there will be a 'cleansing flight': a small crowd of bees circling round the hive and voiding the excreta which they never, if healthy, discharge indoors. As spring draws near and sunny hours cause the crocuses to expand, a new season of activity for bees begins.

In some seasons the queen begins to lay a few eggs soon after the turn of the days. This first effort is rarely more than a small patch, about one inch across, but once well started, breeding increases steadily, subject to suitable weather for foraging; supplies of pollen are collected from early flowers, especially hazel, alder and elm trees. Water is also brought in freely to dilute the honey stored in the comb.

In some years spring growth is steady and continuous and colonies may reach the swarming stage by the end of April; often a cold spell in that month will then keep things back a few weeks, but by the end of May the peak of brood rearing is reached and, unless the bee-keeper can prevent it, swarming becomes prevalent about the time fruit blossom falls and honey income diminishes until the clovers bloom.

3
Hives and Equipment

Although I have given only a mere outline of bee life, it will be sufficient, I think, to give the novice a working knowledge of the subject and I can now pass on to the practical side of bee-keeping, beginning with the most important matter of hives or homes for bees.

Probably because bees *can* be kept in any kind of receptable, varieties of hives have become very numerous and puzzling to the beginner, so let us try and get to the root of the matter.

The essential part of a hive is the comb built by the bees from their own secretions; everything else is subsidiary. As we have seen, bees will build it in the open if unable to find a covered place and several tropical species do this normally. If not cramped in any way, the nest is round, or rather a vertical ellipse, but in any enclosed space it will ultimately take the shape of the chamber. In a hollow tree it will be rounded at the top and extend downwards as a cylinder, but between the inner and outer walls of a timber house—not an unusual place—the combs are long and narrow. I have more than once removed from such a place combs three feet long and only four inches wide.

In the natural way bees attach the comb to the walls of the dwelling, whatever it may be, and if we are content to have fixed comb hives, it follows that any kind of waterproof box or basket will serve as a beehive. However well this may suit the bees, it is very inconvenient for the bee-keeper and the advantages of movable combs are so great that it is now considered essential for each to be enclosed in a frame. Huber's leaf hive consisted of nothing more than frames hinged to each other, so that when closed, they shut the nest in.

FRAMES

The only satisfactory shape for a frame is rectangular; others have been tried, but without permanent success. All must be the same size, and the approved plan is to make them with projecting ends or 'lugs', which rest on a ledge inside the walls of the hive. The inside width of the hive must be half an inch more than the frame, neither more nor less. Amateur hive makers are apt to go sadly wrong in this matter, either by making the hive only just large enough to take the frames, or, on the other hand, leaving as much as an inch of space each side of the frame. In either case, the frames can only be removed by force, and often this causes serious damage to the comb and bees. So long as the space on either side of the frame is between $\frac{3}{16}$ and $\frac{5}{16}$ in, it will be left open as a passageway of great convenience to the bees and equally so for the bee-keeper. I would stress the former point, because ability to pass round the comb at any part must save them much time.

Up to this point all bee-keepers are agreed, but whether the frames should be square, wider than deep, or deeper than wide, or of what size are disputed matters. At least fifteen different sizes are in use, ranging from $11\frac{1}{4}$ in $\times$ $8\frac{1}{2}$ in (Indian) to $19\frac{1}{8}$ in $\times$ 11 in (Quinby), but in Britain only three sizes of brood frame are generally in use. By far the larger number of British bee-keepers use the 14 in $\times 8\frac{1}{2}$ in British Standard frame, chosen by the British Bee-keepers' Association many years ago, but many of the commercial bee-keepers, chiefly those who favour the prolific Italian bee, use the Modified Dadant $17\frac{5}{8}$ in $\times$ $11\frac{1}{4}$ in. Some use the British Commercial 16 in $\times$ 10 in or Langstroth $17\frac{5}{8}$ in. $\times$ $9\frac{1}{8}$ in.

Since in Britain it is most likely that the bees the beginner buys will be on British Standard (B.S.) frames, it is wisest for him to choose that size. The next point to consider is the best number of frames to make up a brood-chamber. For many years ten were considered adequate, but either because queens have become more prolific, or because it is now recognized that swarming is best controlled by giving ample breeding space, many insist that the chamber should have at least fifteen frames. A hive that size is cumbersome and most B.S. users have a ten- or eleven-frame brood-box, so that they can give more room to larger colonies by adding another similar box above, which provides the queen with a run of twenty combs. A modification of this is to use a shallow frame box, as usually employed for surplus honey, which holds frames 14 in $\times$ $5\frac{1}{2}$ in, so that the queen can extend her brood into it if necessary.

A matter of wide divergence is the length of the lugs which project at the top part of the frame. In Britain these have always been $1\frac{1}{2}$ in long, and this necessitates either the addition of a ledge outside the top of the brood-chamber—the usual plan in double-walled hives—or a double wall, as used in the National hive. American bee-keepers have frames with short lugs of $\frac{9}{16}$ in, and this method enables them to be accommodated by a rebate cut in the hive wall, a plan which appreciably reduces the cost of the hive.

Apart from these questions of size and number of frames, other debatable points concern convenience of manipulation. One school of thought, for instance, considers the double-walled hive made of comparatively light wood preferable to single-walled heavier ones, while single-wall advocates point to the time occupied in removing the extra parts of a double-walled hive. When the merits of any of these different styles are considered, always bear in mind that most people prefer and make better use of a tool to which they are accustomed, and are often prejudiced against other styles for that reason alone.

Five distinct types of hives are to-day in general use and it will be sufficient introduction to the subject if I describe and illustrate them.

W.B.C. HIVE

Named after its inventors, W. B. Carr, this hive became widely popular in Britain, having been strongly recommended by the British Bee-keepers' Association.

It has the floor on legs, light inner brood-chamber and outer cases, which leave a space of 2 in or more round the brood-box. It permits shallow frame racks, standard brood-chambers or section racks to be added, either above or below the brood-nest, but the original inventor's claim that the double walls provide insulation to keep the bees warmer in winter, is disputed, for a single wall $\frac{3}{4}$ in thick gives better protection than two half-inch walls, unless the intermural space is airlocked, which it never is. It may be granted that it keeps the bees cooler in summer, but in the British climate that is on the whole a trifling consideration in normal weather conditions.

A considerable drawback is the large number of separate parts which make up this hive, so that the outer cases have to be removed before the working parts are reached. The hive has many varieties, the best being that in which the outer cases are square and overlap each other. The worst are those which *look* square but are, in fact, an inch or so larger one way. They are most irritating and should be avoided. The square form holds ten frames, but some patterns take twelve or more.

Apart from the extra trouble of manipulating, the W.B.C. is a good, serviceable hive, and has the advantage that the inner boxes are light, portable and cheap. Perhaps its chief merit is that it is easier to keep the brood-chamber dry in winter than in other forms, because moisture from the bees can pass through the walls—if unpainted—into the intermural space. All operations can be carried out readily. Owing to the large size of the outer cases and the extra work involved in construction, a complete W.B.C hive costs at least twice as much as a single-walled model.

NATIONAL HIVE

Originally introduced as the 'Simplicity' hive, this form gained in favour since it was sponsored by the Ministry of Agriculture in Britain as the best type of single-walled hive. Its chief merit is that each storey is complete in one square piece, which can be used in any position on any hive of its kind. Its walls are thicker than those of the previous hive and it gives excellent winter shelter. The roof is flat and covered with zinc or felt, is leak-proof and seldom blows off—two things which cannot be said of gable roofs. It can be had in two forms, one shallow, the other deep enough to cover most of the brood-chamber and so give double protection. Eleven frames are contained in the brood-chamber. Perhaps its weak point is the floor, which slopes gently from the back to the end of the alighting place. It does not throw rain off fast enough and damp may seep back into the hive. In some models, little or none of the floor projects in front, so that rain has no lodging place, and the hive is easier to transport on a vehicle. No legs are fitted to this hive, which should stand on bricks or a wooden stage. A Modified National design is now in current use in which the rebates for frame lugs are formed in added side fillets which also provide convenient hand holds.

Single-walled British Modified National hive, showing entrance block, brood-chamber and honey super with shallow roof

MODIFIED DADANT

In Britain this was the most generally approved of the American forms of hive, and is used by several of the larger bee-men. They claim that its large comb area simplifies the inspection of brood in a heavily populated hive, such as would have two brood-chambers of B.S. frames. Its simple case makes it relatively compact and cheap to make. In some respects it is similar to the Smith hive.

SMITH HIVE

A hive commonly used in Scotland known as the Smith, after its maker, is a modification of the National, using B.S. frames, but with shortened lugs as in the American forms. It is correspondingly cheap and has the advantage that those at present using any form of hive with B.S. combs can adopt it without sacrificing any equipment, all that is necessary being to cut down the lugs to the $\frac{9}{16}$ in length.

LANGSTROTH HIVE

Langstroth hive with entrance ramp; its crownboard is fitted with a large perspex window

W.B.C. metal end

The Langstroth hive is a single-walled hive of American origin, constructed on lines similar to the Smith hive but with larger dimensions. Internally it is $18\frac{1}{4}$ in long and $14\frac{1}{2}$ in wide, and the end walls are rebated at the top $\frac{7}{16}$ in into the inner face and $\frac{7}{8}$ in deep to take runners and frames (with short lugs) which measure 14 in $\times 8\frac{1}{2}$ in. It is normally constructed of $\frac{7}{8}$ in timber to give external dimensions of 20 in $\times 16\frac{1}{4}$ in wide and $9\frac{9}{16}$ in deep.

All the English-speaking countries of the world use the Langstroth as their standard hive, though it is not so widely used in Europe and the United Kingdom. There are, however, a number of enthusiastic users who appreciate the advantages of its simplicity and slightly larger capacity.

SPACING

In all hives the frames hang parallel with each other and the distance from centre to centre is $1\frac{9}{20}$ in, so that when the comb is built out there will be an exact bee-space between combs and no room for other comb. Proper spacing is secured by various devices, of which the most usual in Britain are the 'W.B.C. metal ends', which slide on to the lug. They have drawbacks, the worst being that when combs are put into the honey extractor, they are apt to catch in the metal-work, so that they should be removed and replaced afterwards. Short lug frames have no room for metal ends and, in the US, spacing is mostly contrived by the Hoffman type of frame, in which the side bars have their upper third widened so that they are automatically spaced when brought together. These frames are being adopted extensively and may in time oust the old metal end. In addition to the frames, room is allowed in the brood-chamber for a division board. It is not essential, but it is useful when the chamber is very full, for it can be removed easily and then allows room for the frames to be separated. It can also be employed to reduce the area of a brood-chamber when only a few combs are occupied by bees.

COVERING THE FRAMES

To complete the hive, the frames must be covered with something readily removable. This was originally a board, but it was found that the bees stuck this so firmly to the frames that it was hard to remove and quilts were introduced to overcome the difficulty. They are pieces of cloth large enough to cover the brood-box; the one next to the frames must be of tough material to resist as much as possible the efforts of the bees to bite through. Unbleached calico, sail cloth or ticking are good material. The upper quilting should be of a porous nature.

During the last few years, cloth quilts have been much criticized. Besides eating holes in them, the bees stick them down to the frames, making it necessary to pull them off forcibly. The frame tops get dirty with gum and unpleasant to handle. In windy weather, quilts removed from a hive are apt to blow about and be missing when needed. They also need cuts in the middle to permit feeding, and this makes it hard to fit them snugly when the feeder is removed.

Instead of quilts, an improved wooden cover is again being used. This is made of any suitable material—thin board, or the composition board used for panelling and the like. If the hive is made with a bee-space between the frame tops and the top of the wall, the cover can be laid over as it is, but when the frames are flush with the hive top, a $\frac{1}{4}$-in thick rim is fixed round the board so that only the hive wall comes in contact with it. In either case, this leaves a bee-space over the frames, which are kept clean, and room is allowed for bees to pass freely over them—a great advantage in winter. The board is easily removed by inserting the hive tool under one corner. For feeding purposes, a hole about 3 in across is cut in the centre of the board; when not in use, this can be covered by a piece of glass or perforated zinc.

SUPERS

In addition to the parts already described, supers for surplus honey are needed. They are of two kinds, one for the storage of honey for extracting, the other for sections. The former are made in the same manner as the brood-chamber, the size varying with the type of hive. Most people use a shallow frame super, which holds frames $5\frac{1}{2}$ in deep, but many think it a mistake to have two kinds of frame and use B.S. chambers as supers.

Section rack

The rack for sections is a bottomless box, fitted with bars beneath to support the sections, the number of which varies with the size of the brood-chamber. In the W.B.C. there are twenty-one in rows of three; in the National thirty-two in rows of four. In W.B.C. hives, each super must have a 'lift' or outer case to correspond. The minimum requirement is two supers per hive. In a good season, three, four or even more may be required, but in practice, when there are several hives, there are usually some which do not need more than one, leaving spares for the more advanced colonies.

MAKING A HIVE

Hives can be made by any good carpenter, but two points must always be remembered. First, that measurements *inside* the brood-box must be exact, for perpetual trouble will be caused if the bee-space is too large or small. Secondly, as hives must stand not only the extremes of weather, but continual evaporation of moisture from the bees, the wood must be sound and joints well made, otherwise warping and splitting will occur.

Diagrams for making all standard hives can be somewhat difficult to obtain, so that for the convenience of those who wish to make up a temporary hive, I give here the dimensions of a W.B.C. brood-chamber. The material required is:

W.B.C. brood-chamber

1 Two pieces $17\frac{1}{8}'' \times 8\frac{7}{8}'' \times \frac{3}{4}''$
2 Two pieces $17\frac{1}{8}'' \times 8\frac{1}{8}'' \times \frac{3}{4}''$
3 Two pieces $18\frac{1}{4}'' \times 1\frac{1}{2}'' \times \frac{1}{4}''$
4 Two pieces $16\frac{7}{8}'' \times \frac{3}{4}'' \times \frac{1}{2}''$

33

The first two pieces must have rebates $\frac{3}{4}$ in wide and $\frac{1}{4}$ in deep at $\frac{1}{2}$ in from each end. These receive the ends of the narrower boards (2), which must be set flush with the others on one side, leaving $\frac{3}{4}$ in on the upper side. The third pieces are nailed to the ends of (1) on the upper side, care being taken to get the frame quite square. The fourth pieces are used to fill the space between Nos (2) and (3) and should be flush with the bottom edge of (3).

To adapt this brood-chamber as an indoor hive is easy. A floor is made of wood at least $\frac{1}{2}$ in thick. It should be 18 in square and can be made of any width boards, fixed to a couple of battens. On the other side, which will be the upper, nail $\frac{3}{8}$-in boards, 2 in wide, round three sides. The brood-chamber rests on these, so that one side—it does not matter which—has a $\frac{3}{8}$-in opening 14 in wide. This forms an entrance nearly the full width of the chamber, and a strip of wood, say 1 in square, can be used to close it entirely. On one side of the strip, a piece $\frac{3}{8}$ in deep and an inch or so wide is cut out, and when the strip is put with this opening downward a narrow entrance is provided.

MATERIALS AND FINISH

The material of which hives are made must be sound and well seasoned. Lightness is also an advantage. Yellow deal was for a long time considered best, but of recent years Canadian Red Cedar has become popular. It is light, durable and needs no treatment to preserve it from weather, which merely turns it to a pleasing grey shade. If other woods are used, some form of treatment will be needed. At one time lead paint was always used, partly because it is the standard preservative for wood and partly because the colour of the hive has some bearing on the bees' welfare. White paint is a general favourite, but has the drawback of looking shabby very quickly. The other chief favourites are pale blue and stone colour, though green is sometimes chosen. Double-walled hives like the W.B.C. are best painted, but single-walled ones are apt to blister, because moisture from the bees soaks into the wood, and cannot pass through the paint. For these creosote is admirable, for it gives protection from rain, while permitting internal moisture to escape. It is also quickly applied, and much cheaper than paint.

Of recent years, especially since timber became so dear, other materials have been tried for hives. Plastics are being used, also metals, including aluminium, as well as a hive made almost entirely of glass. All these novelties are expensive and only in the experimental stage. It remains to be seen whether they will replace wooden hives to any extent.

INDOORS OR OUT?

Although it is the custom to keep beehives out-of-doors in Britain, there is not the least reason why they should not be kept indoors, as they are in Switzerland and elsewhere. A hive can stand in a shed, or barn, even in a room in the house otherwise unoccupied. Fear that the bees may be dangerous is probably why it is not done more frequently The beginner usually starts with one hive, and lest it should cause trouble, puts it in some out-of-the-way corner. Adding extra colonies from time to time, he puts them in similar outdoor hives, but it would be an advantage to make a special shed to contain

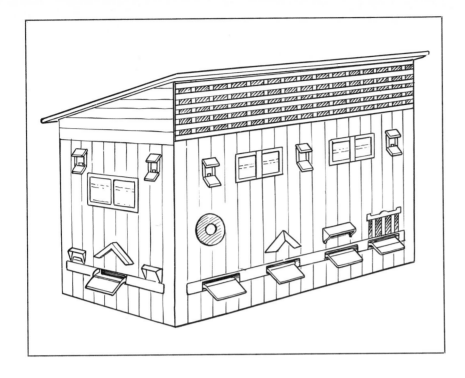

Bee-house, showing varied entrances to enable the bees to distinguish their own hives

as many stocks as one desires to keep. In the long run, this would be the cheaper plan, for the cost of the shed is offset by the saving in the most costly part of outdoor beehives. Light brood-chambers and supers cost very little, and no roofs are needed, for one or two sacks thrown over the wooden cover will provide all the protection required when the hives are under a common roof. One great advantage is that the hives can be opened in windy or wet weather. It is also noticeable that when hives are opened in the comparative gloom of the bee-house, bees do not fly up at the operator, but make for the windows which should be arranged along the front.

Another advantage is compactness—an enormous amount of running about is avoided, especially if the workshop and store are combined with the bee-house, so that the whole business is under one roof.

Furthermore, the vexed question of temperature is almost completely solved, because not only is each hive more protected, but there is mutual assistance in heat maintenance. At the same time there is ample shade in summer when, on very hot days, the bees sometimes come out of an exposed hive *en masse* and hang on the front.

These are not the only advantages. It is easier to arrange the hive platforms at just the right height for convenient manipulation and such work as that entailed in queen rearing, for example, can be done at any time without fear of chilling the brood. Winter examination—to be deprecated in any but urgent circumstances—entails much less risk.

The drawbacks are immobility, which would be awkward for those who go in for any form of migratory bee-keeping, but could be overcome by using one of the lighter portable hives. Danger from fire is almost non-existent in outdoor apiaries, but in a bee-house it is a real risk to be taken into account. It has also been suggested that the danger of disease is greater in a bee-house,

35

but Mr J. Spiller, who is an enthusiast for this method, says that, on the other hand, it is easier to detect disease in a bee-house and thus check it in the early stages.

For those who may contemplate adopting this method, I have drawn a plan of combined bee- and honey-house. The height can be as little as 5 ft in front, but good head room should be provided where the operator stands. Immediately under the eaves, panes of glass or glass substitute should be fixed to run the whole length, with a gap at the top, so that bees can walk straight out to the open. These afford ample light to work by and attract such bees as fly up during manipulation.

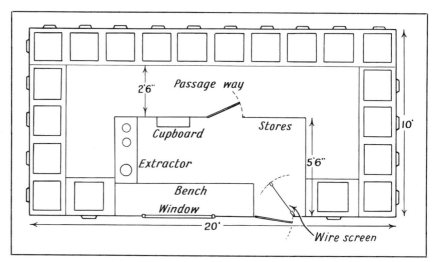

Plan of combined bee- and honey-house

It is desirable that a scheme of colouring should be adopted for the entrances, so that bees can readily distinguish their own hives. They could be painted alternately black, yellow and pale blue.

I have shown this 20 ft × 10 ft house as holding twenty hives, but it is not intended that they should all be occupied. It is really a ten-hive house, but room must always be allowed beside each hive for the work of manipulating or for standing a nucleus beside its parent. If the stand is put at 18 in above the floor, there will be space beneath for storage of supers, etc., thus leaving the extracting room free for purely workshop purposes. The design is no more than a suggestion and can be modified to suit conditions.

OBSERVATION HIVES

Those whose interest in bees makes them wish to study them intimately can have a specially designed hive which enables every detail of the interior to be inspected. Although unsuitable as a permanent home, such a hive can be furnished with a small colony in the spring and maintained all the summer, during which time almost every phase of hive life can be witnessed.

Observation hives are all made so that a comb is accommodated between glass sides. It may hold only one comb, or two or more, one above another. To ensure retention of heat, the glass sides are provided with wooden or cloth covers. The hive can stand in any room, and communicates with the open air through a tunnel leading from the bottom. Furnished with a comb of brood,

36

bees and queen, and kept supplied with honey or syrup, the little hive will enable egg-laying, brood-rearing, etc., to be witnessed in detail. If a comb of bees and eggs without a queen is used, the growth of queen cells and the ultimate emergence of queens can be seen.

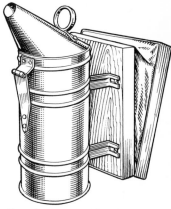

Above: bent nose smoker
Left: portable toolbox. *Rear row:* grass-cutting hook, smoker. *Centre:* smoker fuel, uncapping fork, secateurs, screwdriver, goose-wing bee brush. *Front:* cover cloths, mouse guard, queen cages, hammer and hive tools

TOOLS

The most ancient of the bee-keeper's tools is the smoker. Originally nothing more than a piece of smouldering wood or fabric, the modern article consists of a fire-box connected with a bellows operated with one hand. Two forms are in use. The cheaper is the Bingham, generally considered good enough for the small bee-keeper, but the man who has a lot of hives chooses the pattern which has a larger fire-box and a bent nozzle, much less likely to vomit burning material into the hive. It is easily supplied with fresh fuel and will burn for hours when properly charged. Cartridges of corrugated paper are sold to burn in either, but they are not worth buying; old rags or rotten sacking are better, but the very best material is well-dried touchwood, which makes a lot of smoke, never bursts into flame, and burns away very slowly.

VEIL

Although expert bee-men demonstrating the handling of bees at agricultural shows and the like, do so without using a veil, the beginner will be ill-advised to do so. It is very painful to be stung round the eyes, nose and mouth, and if one is assured that this cannot happen, more confidence is felt when hives are being examined. It is a fact that large commercial bee-men do not disdain to use veils, though quite inured to stings, and even the experts who exhibit bees so calmly without protection will be glad to wear a veil when bees are truculent.

Veils are made of black net or wire cloth, and they are quite easy to make. One-and-a-half yards of fine black net is made into a bag which can be put over any hat, preferably one with a broad stiff brim to keep the veil well away from the face. Black is far the best, giving much clearer vision than white

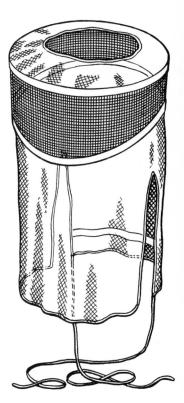

Veil with wire front

37

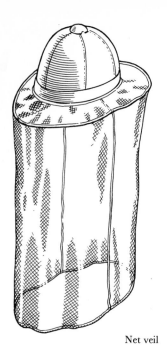

or green. A false hem is made round the open end and a piece of elastic is threaded through it so that the middle portion will go round the back of the neck. The ends are brought through about 6 in apart in front. They must be long enough to pass under the armpits and be provided with hook and eye, to fasten them together behind the back. This ensures perfect comfort and security, and is much better than tucking loose ends into the collar. They can easily come out, and to have bees buzzing inside a veil is worse than not having one. White coveralls make a very suitable dress for either sex, as they can be washed to remove stains of honey, etc.

GLOVES

Those who are very sensitive to stings can wear gloves, but they must be stout gauntleted ones—light fabric is useless. Most people soon get used to a few stings on the hands, and gloves are clumsy things to work in. It has always been regarded as a little bit 'sissy' to wear gloves but, apart from stings, there is another danger to be guarded against. For some reason, when bees are much handled with bare hands, the tips round the nails often become affected, first hardening and then sloughing off, leaving them red and sore. It is believed that some kinds of propolis contain a poison which causes this. It does not seem to affect everybody, but I have suffered from it several times and others complain of the same thing. Rubber gloves would give protection against this, though they are not necessarily sting-proof.

Net veil

HIVE TOOL AND SCRAPER

Hive tool

A hive tool for prising off covers and supers and loosening frames is another essential. It is made of steel and, besides its use as a lever, it forms a screwdriver at one end and at the other is widened and curved into a scraper. Failing this, a stout firmer chisel or screwdriver will serve most purposes. A scraper such as used by paper-hangers is also valuable for cleaning up covers, hive interiors, etc., as well as being useful in handling honeycomb. Other appliances for special purposes will be described in the chapters concerned.

FRAMES

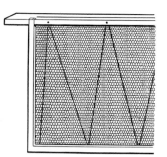

Frame with Lee's wired foundation

I have, I think, made it clear that bees are induced to build their combs in removable frames which hang side by side in the hive. These frames are turned out by machinery in large numbers. They are made with a minimum of four pieces: one long top bar not less than $\frac{3}{8}$ in thick, a shorter bottom bar $\frac{1}{4}$ in. This is sometimes made in two pieces to allow the wax foundation to rest between them. Two sidebars join the top and bottom bars by means of square dovetails. They are cut with accuracy and if driven home fully, ensure a perfectly true shape. They can be bought made up with foundation fitted ready for use, but it is cheaper to buy them in the flat and fit them together at home. After joining up properly, each joint must be secured with a fine wire nail.

All sizes of frame are made in many varieties. The most general is that designed for use with wired foundation, and has a much thicker top bar, rebated on one side to take the foundation, which is held in place by a heavy wedge nailed on.

WAX FOUNDATION

Frames are of no use without wax foundation, because it is the presence of this wax in the frame which ensures that the bees will build inside the frame and not in any direction their fancy leads them. It is sufficient that a small strip of foundation runs along the centre of the frame at top, for from this the bees will build straight down, providing the frame hangs in true perpendicular; but, for two reasons, it is unwise to rely on this. First, if the hive is a fraction out of vertical, the comb will not be centred truly in the frame, but more particularly because, once they are clear of the starting strip, they are likely, if there is a good honey-flow, to make the large drone cells instead of those for workers. Such combs are of no value in the brood-chamber, where we require worker bees to be bred. These two objects can only be secured by fitting a sheet of foundation, stamped with the base of worker cells, completely filling the frame. Foundation is sold in sheets of the correct size for this.

Nor is it sufficient merely to fasten the sheet of foundation to the top bar. It is sometimes done, but the danger is great that a comb filled with brood and honey may, when taken from the hive in warm weather, fall right out of the frame if it is not kept continually upright. It will also be liable to collapse in the extractor. To avoid this, two methods can be adopted. Wired foundation may be used. As the name implies, this has wires running from top to bottom of the sheet, embedded in it during the process of manufacture. This foundation needs only to be fixed into the top bar by nails, the lower end resting between the two bottom bars. Although easy to fix and quite satisfactory when the comb is built, wired foundation is sometimes disliked by the bees, who eat away the wax round the wires, as though trying to get rid of the wire.

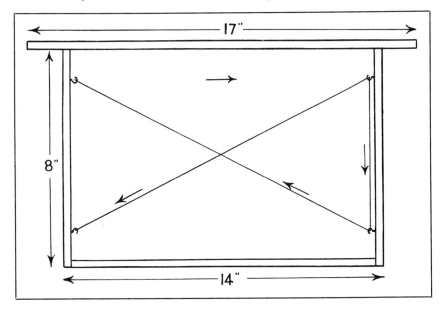

Method of wiring standard frame with diagonal wires

Many bee-keepers prefer to wire the frame and use plain foundation. Opinions differ somewhat as to which is the best form of wiring. Some consider two horizontal wires dividing the sheet into three equal parts the best; others prefer to cross these wires at the centre, as shown in the diagram. I do not think it is very important, but my own fancy is for the latter method. In

39

either case one length of wire is used. Tinned 28 or 30 wire gauge is suitable, though monel rustless wire is now being used a good deal. Before fitting the frame together, two holes are made with a bradawl in each side bar, the wire threaded through all four and drawn quite tight. As it is liable to cut into the wood and so slacken, metal eyelets should first be inserted in the holes. Alternatively, four wire staples, such as are now commonly used for fastening papers together, may be driven in just below the upper holes and above the lower ones. The best way to secure the wires is to drive a tack in just below the top staple and just above the lower one, twisting the ends of the wire round them. Instead of boring holes, some prefer to screw small hooks inside the frame, as in the diagram.

The next thing is to fix the foundation to the top bar. The frames with a stout wedge merely have the sheet of wax laid flush up to the end of the rebate and the wedge nailed on. Those with a saw-cut are a little more trouble, as the cut must be opened to allow the sheet to enter. The best way is to drive two 1-in wire nails into the bench, about 1 in apart, leaving them projecting about $\frac{1}{4}$ in. The heads should be filed off. Stand the frame upside down so that these nails are in the centre of the groove, and pull one end of the frame gently. The groove will open, so that the wax sheet can be slipped in easily.

EMBEDDING

Spur embedder

Now the wire must be embedded into the wax, and for this a wiring board is needed. This is merely a piece of wood half an inch thick, fitting exactly inside the frame. It is laid on the bench and the wired frame fitted over it, the wax *below* the wire.

A special tool called a 'spur embedder'—a toothed and grooved wheel—is heated over a lamp or gas jet and run along the wires. It melts the wax sufficiently to allow the wire to sink in. In default of the proper tool, a bradawl with a groove filed in the end can be used, but it does not make such a neat job.

Those who have access to electric current can embed the wires more quickly and evenly by the current, which must be suitably reduced, or it would overheat the wire. An outfit can be purchased, or those accustomed to dabble in electricity can make a resistance coil. Contact is made with the tacks at one end of the frame round which the wire is wrapped. One can buy a special device which will embed one wire at a time, using a 6-volt battery.

SHALLOW FRAMES

These are fitted up in the same way, except that it is not necessary to use more than one wire, which should run straight across the middle. Fixing the metal ends completes the work. They should go on with the *closed* side inwards, so that they fit up to the side bars. The preparation of sections is described in Chapter 12.

Every bee-man should be prepared to do repairs to hives and other apparatus, as well as to improvise makeshifts on occasion, so he should have a simple kit of wood-working tools. In a bumper season, for instance, it may be necessary to fix up an extra super at short notice, or make a temporary hive for a swarm. In these days of expensive timber, the bee-keeper should cherish any likely bits of wood which come his way.

HIVE SITES

Before setting up a hive, choose a suitable place for it. This is a matter on which it is easy to go wrong from lack of knowledge and cause a good deal of annoyance. It is important to select the place carefully at first, as bees cannot readily be moved during the active season. Two things have to be considered—the comfort of the bees and the convenience, not to say safety, of neighbours and members of your own family. It is desirable to have the hives as handy as possible, and it is quite easy to keep them in the smallest garden if suitable precautions are taken.

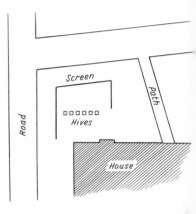

Hives should be sited carefully with screening as shown in the diagrams above, for convenience and safety

Bees always fly direct to a point they wish to reach, so that if there is a field of clover a couple of hundred yards from the hive and nothing in between, they will keep low down. If there is a wall or fence between, they will rise just high enough to clear the obstacle and no more. Generally speaking, bees in a garden surrounded by houses will never be in people's way, because, to reach their foraging ground, they have to rise high enough to clear the buildings, but a garden on the roadside in open country is not suitable, unless a screen high enough to divert the bee-line over the heads of men and horses is placed between them and the road.

When there is any doubt, or experience shows that passers are continually going through the bees' line, the problem can be solved by the erection of screens enclosing the hives in as much space as possible. These screens should be placed as shown in the plan and 6 to 8 ft is high enough for them. What they are made of matters hardly at all, but in a garden a trellis or rustic screen with climbers growing over it is ornamental as well as serving the purpose. Jerusalem artichokes make a splendid screen in summer, when bees are flying most freely and runner beans on sticks or string will also make a good temporary screen. Wire or string netting with mesh not larger than an inch will also serve, for bees rarely attempt to fly through such a small space.

In a garden one has also to consider how far the necessary cultural work will be affected. To stand the bees in front of a plot where weeding and so on must be done constantly will be sure to cause annoyance, so the ground in front should be planted with something like potatoes or fruit bushes which will need only occasional attention, and this may be given in the evening or early morning when there is not much traffic from the hives. In the country, where land is more ample, a favourite place is an orchard, and from every point of view this could not be bettered.

Hives should always be placed so that they are sheltered from strong winds. As these usually come from north-west to south-west in summer, the best site is a south slope. Avoid especially any gap between buildings, where there is likely to be a strong draught. If there is a copse or similar shelter on the north or west side, so much the better. In very exposed places, straw bales can be set up on the windward sides. The object is to ensure that, as far as possible, the air immediately about the hives is calm, for just as ships run most risk when approaching harbour in rough weather, so bees are most likely to be blown down when they are trying to make a landing. The entrance of the hive should face south or south-east, though this is not a very important point, and where many hives are kept it is sometimes desirable to have them facing in different directions.

Hives may be placed close together, or at a distance. In a small garden, it is almost unavoidable to have them close together, but they should be far enough apart—say 3 or 4 ft—to enable one to be attended to without interfering with the others. Where space permits, more room should be allowed. I have seen a hundred hives in one long row with a hedge behind and a good path between hives and hedge, but such a place is rarely available and hives will usually have to be scattered about a field. They should then be 6 ft apart in the rows and the rows 10 to 12 ft apart, while those in one row come between those in the rows before and behind.

Many consider, with reason, that any geometrical plan should be avoided, especially if hives are all of one pattern, bees being then more liable to enter the wrong hive. It is notable that when placed in a long row, hives at the leeward end are stronger than those to windward, owing to drifting of bees along the line. Although the occupants of a hive do not usually permit strange bees to enter, they do not object to any who are laden with nectar and pollen. It is desirable that each colony should keep its maximum strength, and drifting should be discouraged as far as possible, especially because this is a potent way to spread disease through an apiary if one or more stocks are affected. Painting the flight boards in different colours, or putting a large white stone against a hive is a help to bees coming home.

FORAGING RADIUS

Bees have a fairly extensive range of flight, which depends on the amount of food available. In times of scarcity they have been known to travel two or more miles, but this is hardly an effective foraging radius. There should be good crops within at most a mile if substantial surplus is hoped for, and it is well known that bees moved during the active season to any point within two miles of their home, will come back to their old stand in numbers varying with the distance, the older ones having memorized a larger area than the younger. If there are many hives, it is desirable to spread them as much as possible, so that the foraging radius is extended. Even in good bee country, fifty hives is the largest number which should be kept on one spot, and those who keep more, rent 'out-apiaries' in places at least two miles apart, so that the foraging area is utilized to its fullest advantage.

HIVE STANDS

A not unimportant matter is the ground on which the hives stand. Nothing looks better than grass, if it is kept reasonably cut. Anyone who has visited Buckfast Abbey in Devon, England, where the hives stand on squares of concrete in the middle of a smooth-shaven lawn, will appreciate that this is a pleasing arrangement, but unless the mowing can be done very early in the morning or late evening, there will be a lot of trouble for the mower. If the hives stand directly on the grass, there will also be a lot of handclipping to do, so it is an advantage to take out a square of earth about 3 ft each way and fill this in with concrete, or tarmac, flush with the surrounding turf. This will, at least, do away with handclipping.

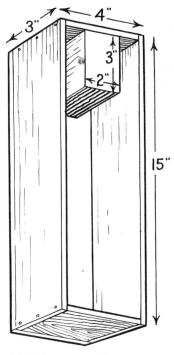

Mould for concrete pillars

Hives with legs can stand directly on such a platform, otherwise a brick or slate should be put underneath each leg, or it will soon rot. Langstroth hives can be supported on bricks or wooden joists and for many years I have kept them on stands made of 2 in × 2 in joitsts, joined together by cross-pieces at the ends. These are laid down on bricks and are long enough to take two hives. This is a very convenient arrangement, for some of the best swarm control methods and those connected with queen rearing, make it necessary for two hives to stand side by side for a certain time. Not more than two should be on a stand, for this enables each hive to be manipulated from the back or side and affords space to put any needful apparatus out of the way.

If hives are kept in a low-lying spot, it is important to make certain that they cannot be flooded. Stands of sufficient height to keep the hives above maximum flood level must be erected and securely embedded, so that they cannot be washed away.

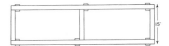

Plan of hive rails on pillars

A good method of supporting hive stands is to make concrete pillars with slots at the top to hold the rails. They are easily made in a mould. A box 15 in long—longer if desirable—3 in deep and 4 in wide is lightly nailed together, and at one end a block of wood 3 in × 2 in is nailed as shown in the diagram. The concrete should be fairly rich, made moderately wet and pressed firmly into the mould. When this is half full, two pieces of stout wire should be put in to reinforce the cement, and the mould filled up. When set hard, the block can be turned out. Four of these blocks will support a pair of hives on rails. Hives or their stands should be levelled carefully with the spirit level, but it is as well to have the back a shade higher than the front, as this prevents any water falling on the flight board from running back into the hive.

PROTECTING THE HIVES

It scarcely needs saying that care should be taken to see that the hives are protected against mischievous interference. Boys of a certain age may dare each other to throw stones or clods at them and may well rouse the bees to fury, with consequent injury to innocent bystanders. What steps can be taken to guard against this depends on the situation. Where stock-horses, cows, goats, etc., are kept, the hives must be securely fenced off, so that there is no danger of their being overturned. Animals should never be fastened up within short range of beehives, for if they happen to be objectionable to the bees, they may be severely, even fatally, stung.

43

4
Getting-and
Keeping-Your Bees

There are several ways of making a start in bee-keeping. Those who already know something of the subject, or have bee-keeping neighbours, frequently begin with a swarm, often presented free at the end of the season, or even found at large. Those who start without such advantages should study the subject well and master the main principles in theory. Then, having obtained the hive and the essential tools described, a colony of bees on six to ten combs can be bought about April. Although this will mean the largest cash outlay, it is the most promising investment, for if the season proves a good one there may be a good crop of honey, perhaps enough to pay for the whole outlay. Such a colony is complete with queen, workers, comb, honey and brood. A few minutes spent in examining it will teach the novice more than a great deal of reading.

Colonies are sent out in travelling boxes, in which the combs are kept in place by a lid of perforated zinc in a frame, screwed down to the side of the box. On arrival, it should be carried to the place where the hive awaits it. If the weather is bad, or it is not at the moment convenient to transfer the bees to the hive, stand the box close to the hive, and open the zinc cover before the entrance hole in front, so that the bees are free to come and go. Something should be put over the box to keep out rain, and it can be left till a suitable day for operating.

TRANSFERRING FROM THE TRAVELLING BOX

Travelling box

The middle of a nice warm day is the most propitious. First, light the smoker and see that it is working properly. On no account put off doing this until the veil has been donned. At least one fatal accident has been caused by setting fire to the veil while lighting the smoker. If you do not wear gauntlet gloves, either fasten the coat cuffs closely, so that bees cannot crawl up the sleeves, or better still discard your coat and wear a pullover. It is also wise to wear clips over the trousers; nothing is more likely to cause panic than to have bees crawling up your legs. This is hardly likely to happen on this occasion, but it is as well to get into the habit of protecting yourself properly, as there are times when bees fall to the ground and climb up the handiest object, generally the legs of the operator.

Now take out the screws which fasten the cover of the box, puff a little smoke through the perforations and, after a few moments, take the lid off gently.

The combs may now be lifted out, starting with the outside one, which should be put in the hive in the same position as it occupied in the box, the others being taken in turn and put close up against each other. Each comb is lifted by gripping the projecting ends firmly and raising it steadily and gently till it is clear of the box. If lifted till level with the face, one side can be examined. To see the other side, turn it in the direction shown in the diagram, because it is wise to keep it always vertical. If the frame has been wired, it will be safe anyhow, but new combs, unwired and full of honey, may drop right out of the frame if held parallel with the ground. This method of turning combs is quite simple and becomes a habit in time.

EXAMINING THE COMB

Take a good look at each comb as it is removed. The outer ones may have only a few bees and some honey, usually towards the top, but the inner ones should have a brood in the centre. Some will be capped over and should present a dull even surface of yellow or pale-brown colour. Unsealed cells will have white plump curled-up grubs, some filling the cell, others lying in the bottom. There should also be little white eggs in many cells. All these middle combs should also have honey along their tops and extending down the sides. It is as well to see the queen, so that one may be sure she has been put in the hive, but

Below left: transferring live bees from a travelling box into a National hive; a cover cloth has been laid over the combs already transferred
Below: turning a comb

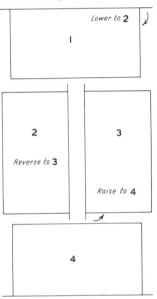

there is no need to be unduly anxious about her. Beginners often fail to find her, especially if she is young and active, but if there are eggs, she is pretty sure to be there. After all the combs have been transferred, any bees remaining in the box should be shaken into the hive by turning the box upside down over it and thumping it vigorously. The cover is put over the frames, the roof over this, and all should be well.

45

If there is plenty of honey in the combs and the weather is fine, nothing further is required at present, but if there seems to be very little food and the weather is not favourable, a feeder should be put on and a good supply of syrup given. Syrup is made by putting two pounds of refined sugar into one pint of boiling water and stirring till melted. Various methods of feeding will be described later on, but a good makeshift feeder is a two-pound jam jar. After filling this with syrup, put two thicknesses of muslin over and tie down, or fasten with a stout rubber band. The jar is turned upside down over the feed-hole and left till emptied, which will be in a day or so, according to the strength of the colony.

It is not always easy to buy a full colony of bees in the spring, since most bee-keepers are disposed to keep them for the crop they are likely to get, and it is usually easier to get a 'nucleus' during the summer. A nucleus consists of three to six combs with bees and queen, which should be a young one of the current season. The outlay is considerably less, but the likelihood of surplus honey being obtained is rather remote, though not unknown in a good season. A nucleus is transferred and treated in the same way as a full colony, but will need feeding until it has attained full strength. As the bees fill up the combs, frames of foundation are added one at a time on the outside of the rest until the brood-box is full.

BUYING A SWARM

Swarms may be purchased from May to July, their price depending on their weight—they are commonly sold by the pound—and the month. The earlier they are the more costly, as a May swarm is likely to gather a good harvest, whereas one arriving at the end of the season will probably have to be fed. A May swarm of 6 lb will give its value in honey in a reasonably good season.

Swarms are sent out in boxes or skeps and have to be hived in a different manner from stocks or nuclei. First, it will be necessary to have prepared frames of foundation. These may be bought all ready for use, or made up as described on pages 39–40. For a swarm of not less than 4 lb, the full complement of ten frames should be used and the brood-chamber filled up with them. A smaller swarm should have less, say six, others being added later as it completes them.

Having filled the brood-chamber and put the cover on it, lift it from the floor and stand it on one side. In its place put an empty brood-chamber, or shallow super. Take the cover off the box or skep containing the swarm. Hold it mouth down over the empty chamber, and jerk the bees into it. Put the prepared brood-chamber over and leave it until the bees have climbed up and settled among the frames. They will do this by evening and the empty chamber can be removed, one person holding the full one up while another takes the empty one away. This plan is better than the old one of throwing the swarm in front of the hive, for it can be done at any time without fear of the bees taking wing. If the old plan is adopted, it should not be carried out till sunset.

Unless there is a heavy honey-flow on, it is wise to feed swarms generously for the first few days.

HANDLING BEES

The fear with which many people regard bees is out of all proportion to the harm they can do. As in learning to swim, the tyro makes rapid progress the moment he loses fear and strikes out boldly; so in bee-keeping, after a few stings have been received and their comparative triviality realized, they cease to be a bugbear. At the same time, it is desirable on all grounds to avoid them, and judicious handling, which becomes instinctive in time, will reduce stinging to a minimum.

A good smoker is the first requisite and, before an operation is started, it should be charged sufficiently to outlast the job. The nozzle should be kept clear, so that a good blast of smoke can be given at any moment and, to keep it alight, the smoker should stand on end when not in use, the nozzle acting as a chimney. In this position it will keep alight till the fuel is exhausted. If laid on its side it goes out quickly. See that the nozzle is secure, or burning fuel may fall into the hive and set the combs alight. The bent nozzle type is far the better article, but even from this burning ashes can be blown into the hive so that it is wise, after it has been well lighted, to put in a little fresh grass. This holds back the burning material and cools the smoke.

Some bee-keepers use a carbolic cloth as a substitute or supplement to the smoker. This is a piece of cloth 18 in square sprinkled with carbolic acid. The standard recipe is one ounce of No. 5 Carbolic to two ounces of water. This mixture is kept in a dropping bottle, or an ordinary medicine bottle with a piece cut out of the side of the cork. The cloth is sprinkled fairly liberally with the mixture, rolled up and kept in an airtight tin. This cloth has the same effect on bees as smoke: they run away from it, plunge their heads into honey cells and fill themselves with honey. In this gorged state they lose all desire to sting and can be handled as harmlessly as flies.

However, the cloth is not much use till the cover has been taken off completely, whereas smoke can be blown in at the hive entrance, or in the first corner of the cover loosened.

It should seldom be necessary to keep the cloth on or to apply smoke continuously. If three minutes are allowed after the first application, so that the bees can fill up, all will be well, but continuous smoking merely drives the bees hither and thither and gives them no opportunity to imbibe.

A hive should always be manipulated from the back or side: never stand in front of the entrance with bees coming and going. If there is a path behind the hives, a good operator can get well started before the bees are disturbed.

CLEARING THE COMB

Combs should be handled and turned only as described on page 45, but to remove bees from them, shaking is the best plan. Holding the ends firmly, the operator lifts the frame vertically over the open hive and jerks it sharply down. An even better plan is to hold the frame by one lug, allowing it to hang down, and then strike the hand holding it with the flat of the other. As a rule, combs should only be shaken over the open hive. Worker bees will find their way back even if they fall outside, but newly hatched bees, and especially the queen, may be unable to do so.

Sometimes, as when there are queen cells it is desired to preserve, it is not safe to shake the combs and it is then better to brush the bees off. A stiff goose wing can be kept for this purpose, but I prefer to use a little bunch of stiff grass stalks. The bees should be brushed downwards, holding the comb by one lug.

OPENING THE HIVE

Before opening a hive, have everything ready for the operation to be performed and be quite sure you know what you are going to do. Hives should not be opened without a definite purpose and that should be carried out without interruption. If it is absolutely necessary to leave in the middle of a job, the cover should be put over the combs and the roof replaced: one can never be sure not to forget or be unable to come back before a shower. On no account should combs be left out of the hive. It might seem needless to say this, but I have known a bee-keeper take out brood combs, go away and forget them, leaving them standing beside the hive for days, with bees clinging faithfully to them. Brood combs should never be left exposed, either to cold or the sun, more than can be avoided, as long exposure to either is harmful.

TEMPER AND TEMPERATURE

If a stock proves so truculent that neither smoke nor carbolic will subdue it, it may be because there is no unsealed honey for the bees to run to. If a little warm syrup is poured over the frames, they will make use of this and become amenable.

Bees behave quite differently at different times. When the weather is warm and genial and nectar is abundant in the fields they are quite amiable and need scarcely any subjugation. On cold windy days, when foraging is at a standstill, they behave very differently, often rushing out to sting the moment the cover is lifted. Thundery weather is also notoriously irritating to them. Such times should not be chosen for opening hives if it can be avoided. In the autumn great care should be taken, not only to subdue the colony before operating, but to prevent robber bees from other hives attacking it. If fighting breaks out when a hive is open, cover it down at once and, if it continues in the air around, get the garden syringe and spray the flying combatants till they go home.

Accidents will sometimes happen. A hive may be overturned, or a comb or super dropped, rousing the bees to fury. If one is properly protected and smoker or carbolic is at hand, things can be put right at once, but if not, the best plan is to throw some sacks or other material over the exposed parts and leave things alone for an hour or so before taking further action.

STINGS AND STOICISM

Needless to say, the bee-keeper, while doing all he can to avoid stings, must bear them bravely if they are received during an operation, carry on and finish the job, or at any rate replace everything taken from the hive and cover up properly before leaving. The effect of stings depends a good deal on the individual constitution, some people suffering more than others. As a rule, swelling increases for twenty-four hours and subsides during the next twenty-four. Local swelling need cause no alarm, though it may be very great. The effect of a sting may be made worse by treatment. Most people try and pull it out, but this presses the poison bag and squeezes more into the wound. The proper way is to *scrape* the sting off with the thumbnail or a knife.

Various things are advocated as treatment. Iodine, ammonia, onion juice, tobacco, and several proprietary remedies, rarely do more than allay irritation,

48

and a little honey smeared on is just as effective. In the case of great swelling, hot fomentation is the most effective treatment. Serious results are not unknown and a few people are so allergic to stings that they are liable to collapse. Medical aid must be sought in such cases. An injection of adrenalin is usually given, but, failing this, a strong stimulant is indicated.

After a few stings have been received, nine people out of ten become more or less immune and all but indifferent to them.

EVERYTHING TO HAND

From what I have said, it is clear that quite a number of things should be on hand when one opens a hive. To make sure they are, I have a box made of plywood. It measures 15 in each way, since that size enables me to catch a swarm in it, hang combs taken out of a hive temporarily in safety, and hold all the things I need when at work—smoker, hive tool, bottle of syrup, dredger of flour (used for uniting colonies), knife, tin containing carbolic cloth, spare frame of foundation or comb, spare smoker fuel, and anything else which the special work may require.

I find it very helpful to have at hand when manipulating stocks, two or three clean burlap bags. One of these can be thrown over an open hive at a moment's notice without injuring bees. If this is done when signs of truculence are noticed, the outbreak is smothered at the start, and in a few moments' time more smoke can be given and the work proceeded with. If a stock is divided into two for any reason, the part not being examined can be covered in the same way, so that it is not at the mercy of robbers. Something may have to be fetched from store, or a slight repair may be necessary. If a supered stock is being examined, the removed supers should be covered in this way and the risk of interference from outside will be reduced to a minimum.

These precautions and quiet, unhurried, methodical working, will make all the difference between a peaceful apiary and one which is a menace to the neighbourhood.

SUPERING

The beginner should not be too disappointed if he fails to secure honey for himself in the first season, though if he starts with a full stock in April, he should certainly get some in an average year. Indeed, it is often noticed that some novices get a particularly good crop from the first stock. This is generally called 'beginner's luck', but it does not always come off. If he gets a good working knowledge of the first principles of the craft and builds up a good stock for the future, the novice may well be content.

A June nucleus on four combs will rarely do more than build up its strength during the main honey-flow, and it will not be necessary to do more than add frames of foundation one by one, until the brood-box is full.

Full stocks and swarms will require 'supering', that is to say, when the brood-chamber is all but full of bees and brood, another chamber is added above it. It will have been noticed at the very first inspection of the combs that bees put their honey at the top, as far as possible from the entrance, or on the combs outside the area in which the queen breeds. It does not always remain where it was first placed. Foragers put it in the empty cells which come handiest, as they are eager to go and get more. It is very thin stuff and it is partly by being moved about that it loses its water content and becomes thick and viscid. The house bees carry away any in the centre combs so that the queen may use the cells for breeding and they naturally take it higher up as the brood spreads out. By putting surplus chambers on top we conform to the bees' habits, at the same time ensuring that surplus shall be in the most convenient place for removal.

SUPERS AND SEASONS

How often supers are needed depends, of course, on the season, which is unpredictable, but it should be a rule never to let bees run short of room. It can do no harm to put a super on before it is actually needed, but to delay doing so when the brood-nest is full will almost certainly cause a swarm, with the inconvenience and sometimes loss which ensue. In the height of a good season, a stock will gather from four to ten pounds of nectar daily and, as a standard comb only holds about five pounds of matured honey, it is obvious that the space available in the brood-chamber for the thin and light nectar coming in will soon be choked. It is a fairly sound rule to put on a super as soon as the first bees begin to appear on the two outer combs of the brood-box.

The first super should consist either of a chamber the same size as the brood-chamber, or a shallow frame super and it must, like the brood-chamber, be fitted with foundation. It is far better to use full sheets, though it is not so important as for the brood-chamber and, in emergency, starting strips may be used. It is not true economy, however, because expensive though foundation is to buy, it is much more expensive for bees to make wax, for at least ten pounds of honey must be consumed to make a pound of wax, and naturally with a good supply of wax already placed in a convenient position, the work of building up the comb proceeds much more rapidly.

USING THE EXCLUDER

A matter on which there has always been sharp division of opinion is the use of 'excluder'. This is a metal screen covering the top of the brood-chamber. It may be made of sheet zinc closely perforated with slots, through which worker bees can just squeeze, but are too small for the queen to negotiate. This is the oldest and by many still considered the best pattern, but a modern form is made of stiff wire. This gives more space for workers to get through and has that much advantage, but there is not a lot to choose between the two patterns.

Some old hands will have nothing to do with queen excluder. They contend that the queen should not be hampered in any way, because the more brood there is, the stronger the colony will become and the more honey it will store. Those who favour excluders complain that not only is brood mixed up with honey if the queen goes into the supers, but that much pollen is also stored there, spoiling the combs for future use.

My own view is that the truth lies between the two extremes. If the brood-chamber is the minimum 10 B.S. it is not sufficient for a queen who is worth her salt, and the first super should be put on without excluder, so that she may, if she wishes, use it for extending the brood. She will rarely use more than four or five of the central combs, and towards the end of the honey-flow, when

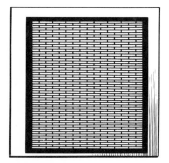

Zinc excluder

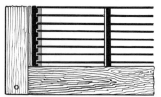

Wire excluder

breeding slackens off, she will usually have gone below and the bees will fill up the cells with honey as they become empty. On the other hand, if a large brood-chamber is used—12-15 B.S. or 11 M.D.—excluder should be used between it and the super.

Unless there is a really strong honey-flow, bees do not go through excluder readily and in a poor season may never do so. They can often be tempted up by a comb of honey put in the centre, and this is one advantage of using standard combs for supers, because a comb of honey can be brought up above the excluder and an empty one put in its place. This starts the storage of honey above the excluder and increases the breeding space below it.

A peep inside the super after it has been on a week may show that it is well advanced, and as soon as it is half-full, another should be put over it, this time with excluder between the two—if there is not one below—as the bees will certainly go up if the lower one is filled. Some bee-men put the second super under the first, but I am confident that it makes no difference in a good season and in a poor one it is much better to have one full super than two half-filled ones, for when a gap is made between the brood and the honey stored, the bees will bring honey from above to fill it, if it is not coming in fast from outside. The main contention is that since the first super will be completed first, it should be at the top for convenience in removal. There is nothing in this, for very few experienced bee-keepers remove honey until the end of the season, when they take it all off at once.

A strong swarm should be treated rather differently from an established stock. The queen has usually passed the time when she lays to her full extent and will not need so much room, while the swarm, consisting mostly of foragers, is strongly bent on honey storing. After feeding it for the first few days, a super should be put on with excluder under it. This may be either shallow super or section rack. The bees will rarely refuse to enter it at once and it is an excellent opportunity to get a few of the much-prized sections.

ADDING A SUPER

Adding a super is a simple matter. Remove the hive roof. If it is a flat one, turn it upside down beside the hive and stand the super on the edges of this improvised tray by turning it at a slight angle. Insert the hive tool under a

Far left: adding a wooden-framed queen excluder
Left: adding a second super; the bees have been lightly smoked off the upper surface of the first super

51

corner of the wooden cover and lever it up gently, puffing in a little smoke to drive the bees down. Lift the cover and put it on top of the super. There will be bees adhering to it, but they will not be crushed if the cover is lowered gently. Now send a waft of smoke across the open brood-chamber and as soon as the bees have run down, lift the super and place it in position. To avoid crushing bees, first put one edge on the corresponding part of the brood-box and then lower gently. Sometimes the bees built pieces of comb on top of the frames: these should be scraped off with the hive tool, the scraping being dropped into the toolbox. If excluder is to be used, it must be laid correctly over the brood-chamber before the super. If it is lifted up and down two or three times, bees on top of the frames will soon get out of the way. A carbolic cloth may be used instead of smoke and this will clear the frame tops speedily and completely.

SWARMING

Bee-keeping would be an easy occupation if colonies properly provided with room to breed and store honey would continue doing it without interruption, but this is just what we cannot be sure of. At any time during the summer the most prosperous and energetic colony may suddenly cease its labours and swarm. The experienced bee-keeper cannot always recognize the imminence of this event by outward signs, so it is no wonder that the novice is taken by surprise, or that sometimes a swarm disappears without his knowing it, so that the harvest of honey he saw being built up turns out to be little more than a mirage.

Swarming, indeed, is the chief problem of bee-keeping. It is not like sting-ing, one which grows less as you become accustomed to it, and many beginners have been disgusted and discouraged from continuing because they knew no way of getting over it.

It is peculiar to the honeybee, for although associated with reproduction, it is not a mating flight, like that taken by ants and termites, the only other social insects which fly out in a body: the virgin queen usually takes her wedding flight unaccompanied, and the swarm has already left the hive with the queen mother. This first or 'prime' swarm is, in fact, a migration and associated, as all migrations are, with the fluctuation of food supply. Closely related species of bee native to India migrate regularly, leaving a district where nectar sources are waning for one where they are waxing. In higher latitudes, where there is prolonged winter with complete absence of flowers, this habit would be fatal; if a colony is to survive, it must store enough to carry it through, instead of consuming all its gains in breeding, as the Indian species do.

JANUARY TO MAY

From January to May there is a gradual increase of flowers and the growth of bee colonies coincides with this. Towards the end of May, spring blossom passes its peak and there is a decline in nectar income, so that there is generally a short period when bees consume as much as they gather daily. It is during this period that prime swarms usually appear and it is not, I think, a coinci-dence, for the slackening of bloom checks the foraging instinct and causes the hive to become more congested. Bees which have reached the stage when they should take part in foraging are unsettled, and it is, I feel sure, this feeling of insecurity which is the mainspring of the swarming impulse.

Swarms which appear before this significant stage are usually 'supersedure' swarms, due to some defect in the queen which causes the bees to replace her. They do not always swarm under these conditions, and when they do the queen is often killed and the swarm returns to its hive.

MAY TO JUNE

Real prime swarms generally appear from mid-May to mid-June, according to the state of the season.

If nothing is done to a stock which has thrown a prime swarm, it will often send out a second swarm or 'cast' when the young queens are ready to fly. This is usually about nine days after the prime swarm issued, but it may happen before if the swarm was delayed. In bad weather, the bees keep mature young queens in their cells till a fine day comes, and it is then not unusual for one or more to leave with the swarm.

Casts, which are headed by a young queen, often fly off unobserved, and some races of bees throw off so many that the stock is seriously depeopled.

REASONS AND SIGNS

Swarming may be advanced or retarded by the growth of the colony or the size of its abode, and it is certain that a stock cramped for room will be more likely to swarm than one in a large apartment. Still, these are not the sole considerations, and when weather conditions favour it, colonies will often swarm regardless of their size and the space at their disposal.

The earliest sign is the appearance of drones. After a while you may notice that the bees are not rushing out of the hive as eagerly as before and there is a tendency to cluster about the entrance. Although this does not necessarily mean imminent swarming, it shows that nectar sources are not at the moment abundant, and if it is intended to adopt either of the measures described further on, this is the time to examine the stocks to see if queen cells have been started. At the same time it is well to repeat that stocks sometimes swarm without showing any sign of slacking off.

GUIDING A SWARM

Whenever possible, the bee-keeper should make a point of being on hand during the middle of the day in the swarming season. If he actually sees a swarm leave the hive, he will be in a much better position to carry out the operations recommended. As it is never certain where it will alight, steps should be taken to guide it. This is comparatively easy if the right material is at hand—a bucket or two of water and a powerful syringe. A garden hose is ideal: by pouring a spray of water over the flying bees, they are induced to settle and, if there is no handy bush or post, will come to the ground. Failing water, handfuls of dust will produce the same result, though not so surely as water.

The old-fashioned idea that swarms will be induced to alight by a tintinnabulation by gong, bell, or some combination of domestic implements, like a tin pan and the key of the back door, is now regarded as a superstition. However, its origin would appear to be rational enough. In days when every country cottager kept bees it was important to ascertain from whose hives a swarm had issued so that the first person to see a swarm flying, beat the gong to inform the neighbours and give them a chance to claim the swarm if they believed it to be from their own hives.

COLLECTING A SWARM

Once the swarm has settled, it should be collected at once. The traditional receptacle is a straw skep, which is roomy and light, no small advantage if the swarm has to be taken from a place only reached from a ladder; but of course, in the old days, it was the permanent home of the swarm, whereas nowadays it will only be temporary and its place can well be taken by any suitable object. I have described the toolbox I use for this and other purposes, but I have on occasion used boxes, buckets, baskets, sacks and even my hat. A very light and convenient receptacle is one of the cardboard cartons used for packing many different commodities. If the top flaps are turned outside and fastened securely together, it will serve admirably.

Hiving a small swarm from the skep in which it was collected into a nucleus box, using a cover cloth and mist spray

A swarm on a bough near the ground is easily taken. Holding the box beneath it, the operator seizes the bough and shakes it vigorously, so that all the bees fall in. The box is then laid on the ground and turned over, a stick being put under one edge to allow bees to go in and out. If the queen is in, the others will soon follow: if not, they will come out and return to the bough.

54

Far left: small swarm on a bough complete with combs and brood
Left: collecting a swarm in a straw skep

The operation must then be repeated. Sometimes shaking is impossible, as on a wall, post or tree trunk, from which it is necessary to sweep them. A bunch of grass, or a stiff feather will serve and a few downward strokes will brush them all in. Instead of turning the box upside down, it is better, if one has them available, to put two or three combs in over the bees. They can then be hived without turning them out again.

The only suitable plan for removing a swarm from a hedge or thick bush is to prop the box over it and drive the bees up with smoke. It is even better to use the brood-chamber of a frame hive. Some frames of foundation or preferably combs, should be put in the chamber and covered with a sack. The brood-chamber is then placed as nearly as possible over the swarm and made secure by one or two props. When the bees have gone up, the brood-chamber can be placed directly in the position it is to occupy. Swarms on the ground are easily hived in the same way.

Swarms in trees give more trouble. Sometimes they are readily accessible from a ladder and so placed that they can be shaken or brushed into the box, but if they are on a slender bough this is not always feasible. It is sometimes possible to take them by fixing a skep or box on the tines of a pitchfork, raising this till the swarm is inside, and striking it sharply against the underside of the bough. Often the best plan is to cut off the bough. Before doing so, a rope should be fastened as near the swarm as possible, then passed over a bough above and brought to the ground, where one person holds it while another cuts the bough. It can then be gently lowered. Whenever a swarm is on slender branches, it is always the best plan to cut off and carry the whole thing intact to the permanent site, where it can be shaken directly into the hive.

OTHER WAYS AND MEANS

Awkward situations will sometimes arise and must be dealt with as circumstances dictate, but swarms can always be induced to move by smoking or feathers dipped in carbolic solution. Care must be taken not to overdo this, or the swarm may take wing and fly right out of reach. Old brood combs are always good bait for swarms and if one or two can be fixed just above or beside a swarm, it will soon cluster round them and can be removed. It is, by the way, useless to tempt them with honey or sugar syrup: being already fully loaded with honey, it has no attraction for them.

When bees were kept in skeps, hiving was complete once the swarm was in, but nowadays they have to be turned out again. I have explained how this is done, but it is much better to use a box that will allow standard combs to be put in over the swarm, which will then climb on to them and can be transferred to the permanent hive at leisure. Otherwise it is not safe to turn a swarm out in front of the hive till evening.

Swarms newly-issued are conspicuously amiable, owing to their gorged condition, and can usually be dealt with without a veil, but one which has been out of the hive in cool wet weather, uses up its rations and becomes particularly touchy. Before tackling one of these derelicts it is wise to spray it well with some thin syrup to sweeten its temper.

VALUE OF SWARMS

Under the skep system of bee-keeping, prime swarms were looked upon as *the* producers of the harvest. The earlier they arrived the better. They were put into a completely empty hive, so the combs they built were quite new, or as they were called 'virgin'. Coming just before the main flow, swarms concentrate on foraging, raise little brood, and in a reasonably good summer, fill the hive with honey.

They were nearly all old bees, the queen was two or more years of age and so few bees were raised in autumn that they stood a poor chance of surviving the winter. Hence nothing was lost by destroying them, the purest honey was taken, and the old stocks with young queens were much better fitted to face winter.

Under the modern system, bee-men are apt to make no distinction between swarms and stocks and often try to winter swarms with old queens, but it is a great mistake. Prime swarms should, by one or other of the plans described in the next section, be made to concentrate on nectar gathering, and when the season for this has passed they should either be requeened or broken up to strengthen nuclei of the current season.

MANAGEMENT OF SWARMS

The excitement and trouble involved in hiving swarms are not the main disadvantages; within reason, it is a pleasant adventure, so long as it is successful, and I think even old hands catch some of the bees' enthusiasm when they see a swarm whirling madly over the apiary. However, it is the effect on the honey crop which is most to be deplored.

When a swarm leaves, the foraging force is split in two and the nectar gathered in proportion to the demands on it is reduced. The swarmed stock is usually much less active, while it waits for its new queen to mate. It generally has a large quantity of brood to attend to and relatively few foragers. The swarm, it is true, concentrates its energy on foraging and if provided with complete combs will often fill them in a few days. Otherwise it would consume a large part of its gains on building comb.

There are several ways of minimizing the ill effects of swarming. Early swarms, say those issuing before the second week of June, can be treated as separate stocks. If your apiary is full, it is not a bad plan to sell such swarms, which make a good price and leave you with a requeened hive, which has every chance of regaining foraging strength before the main honey-flow.

Later swarms will, however, react unfavourably on the honey crop unless suitably dealt with.

The oldest plan is to find the old queen in the swarm, kill her and return the swarm to the parent hive. Supers are added to allow ample storage space and if the honey-flow is on, the stock will devote its energy to filling them. In due time a young queen will be mated to head the colony.

This plan, however, has grave dangers. There are usually several young queens in their cells and when the first is ready, she may leave the hive and take all the bees which swarmed before with her. The swarm may be even bigger and, the queen being young, may go right off without pause. To prevent this it is essential, before returning the swarm, to destroy all queen cells but one. If this is done properly it will prevent a second swarm, but as queen cells will be in all stages from sealed and ripe ones to others hardly started, it is easy to overlook some. To do the work properly, all the bees must be shaken from the combs. The cell retained should be the finest and most mature. A further risk is involved in this method. The virgin queen may be lost on her mating flight, or bad weather may delay her mating beyond the twenty days usually considered the limit of time during which she can be fertilized. She will then be a useless drone breeder and the stock will perish. As a safeguard against this, it is wise to save another queen cell in a nucleus colony. Two or three combs well covered with bees, and one queen cell, are put in an empty hive. The bees from one or two more combs are shaken in with them to make good the loss caused by older bees returning to the old home. If either queen fails to mate, the two lots can be reunited. If both survive, the nucleus can be built up to form a new colony.

This plan is not often used; mainly, I think, because so few people are able to find the queen in a swarm. It is also unnatural, and experience shows that better results are achieved by working *with* the bees and not against their ingrained instincts. The natural plan is for the swarm to settle down as a new unit and this was followed by the old bee-keepers, but since it involves the establishment of two colonies, it obviously means that the surplus available for the bee-keeper's harvest will be less.

Bearing all this in mind, a system has been evolved which enables us to keep all the foraging bees in a single unit and so get little if any less harvest than would have been secured from a non-swarming stock. The basis of this system is the separation of the old queen from the brood, which is precisely what happens when a swarm is thrown off.

PAGDEN'S METHOD

The first step in this direction is what is known as the Pagden method, because it was originally described by J. W. Pagden in a booklet published in 1870. This plan was originally practised with fixed comb hives, but can equally well be used for frame hives. It is simplicity itself. Instead of putting the newly hived swarm on a fresh site, it is put on the stand occupied by the parent stock, which is moved to a new site. The consequence is that the swarm is reinforced by all the bees flying at the time, for they naturally return to the known spot and for at least three days after hiving the swarm becomes stronger. Having been seriously depleted in numbers, the old stock rarely throws a second swarm, but in due course the first queen becomes the new head.

This simple plan has been modified in many ways. Instead of moving the old stock right away, some put it beside the swarm, turning the entrance at right angles from the swarm. It is left thus for three days and then taken away to a fresh site. A further batch of foragers is added to the swarm in this way.

DEMAREE'S METHOD

The same principle of keeping the queen separate from the brood is followed in the American, Demaree's, plan. This can only be applied to frame hives. The swarm is hived, put on the old stand and covered with queen excluder.

Specially enlarged cells in which queens are raised

The supers, if any, on the old hive are put over this, any queen cells in them being destroyed. Another excluder is put over the supers and the old brood-chamber, with brood and queen cells, is placed on top of the lot. All the queen cells may be destroyed if no increase in stock is desired, but as the bees will start fresh ones, an examination must be made in seven days' time and the new cells also destroyed. In this way all the bees are retained in one hive and, given favourable conditions, should store as much honey as an unswarmed stock.

SNELGROVE'S METHOD

A further modification was introduced by L. E. Snelgrove in 1931. He devised a board to cover the chambers of a hive. It is made like the usual wooden cover, but the central hole is permanently covered with wire cloth with a rim round the outer edge, above and below, $\frac{3}{8}$ in deep. This can be from $\frac{1}{2}$ to 1 in wide. On three sides of this rim, pieces are cut out both above and below, about an inch wide, the pieces being retained and fitted with a knob or staple on the outside, so that they can be taken in and out as required. This 'screen board' is put on top of the replaced supers, instead of the second excluder. At first all the lower openings are kept closed, and only one of the upper ones is opened; this enables bees in the top chamber to fly as desired. After five days, this opening is closed and the one below it opened, so that returning bees enter it and so join the bees in the supers. At the same time, the

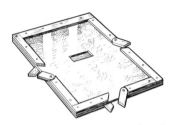

Snelgrove screen board

58

upper opening on the opposite side is opened, so that top chamber bees can leave. Again this is closed and its lower counterpart opened after five days. The third pair of openings is dealt with in the same way. Maturing foragers are thus steadily drafted into the supers, until all brood has emerged, when the board can be removed and the chamber used as a honey super.

Queen cells may be destroyed, as in the Demaree system, or used to make increase, as desired.

Indeed, the board is very useful to those who wish either to make increase or merely to requeen the stock at the end of the season, for the queen cells are conveniently placed for inspection and the bees in the top chamber are young and inoffensive. If necessary, the chamber can be divided into two or more parts and more than one queen can be raised and mated there, or the cells may be caged as soon as they are 'ripe' and transferred to nuclei. There must, of course, be no alteration of the openings after the queen has left her cell and if she is to be mated from this chamber, it is better to have the opening at one side, as she is then less likely to get into the lower portion when she returns.

SHEPHERD'S METHOD

Mr W. Shepherd of Glasgow, Scotland used a modified version, which he claimed to be more effective and easier to operate. He uses a screen board, with only *one* opening, communicating with the top chamber. Bees leaving by this opening, pass down the hive, either through a specially fitted tunnel, or, in a double-walled hive, between the inner and outer cases, to a point level with the hive entrance. When returning they do not go up the tunnel, but straight into the hive entrance. My only criticism of this plan is that too many bees are drawn away from the brood at first and on a cool night some might be exposed and killed.

SWARM PREVENTION

The trouble and risk of serious loss from swarming continually impel bee-keepers to find means of preventing it. As I have said, I think the chief cause, apart from its connection with the reproductive impulse, is a feeling of unsettlement and insecurity. This may be caused by many things, such as lack of breeding space or storage room, overcrowding, insufficient ventilation, want of shade, superabundance of larval food, lack of food, high temperature, impaired vigour in the queen; also intermittent honey-flow, presence of queen cells, or disturbance of the brood-nest. All or any one of these things are likely to cause dissatisfaction with the place and urge the colony to try somewhere else.

What seem to be the more cogent facts are:

1 Certain races of bees (Dutch and Carniolan) are more prone to swarm than others (Black and Italian).
2 Colonies with old queens are more likely to swarm than those with young ones.
3 Stocks in large brood-chambers are less disposed to swarm.

Overriding all these things is the nature of the season. In some years swarming is very prevalent and nothing will stop it. Generally speaking, a fine dry summer has a short swarming period, while in a patchy season swarming is often protracted and tiresome. It is pretty generally agreed that a good honey season is also a swarming season. To keep a race of bees not inclined to excessive swarming, to have colonies headed always by young queens and to provide ample breeding room are, therefore, the first steps to swarm prevention. They will not entirely stop it because, to have good crops of honey, we must induce bees to breed early and freely, so that when the honey-flow comes there will be a greater proportion of adult bees. We are thus obliged to act in a way which predisposes to the congestion which many regard as the main cause of swarming.

There are two main systems of swarm prevention: regular removal of queen cells, and artificial swarming or division of stocks.

The late Dr C. C. Miller was the chief exponent of what he called the 'cell killing' method, now generally known as the 'ten-day' plan. I cannot do better than quote his own description of it.

'We begin looking for queen cells as soon as we think there is danger of their being started. The first time we look in the strongest colonies. If we find no cells started we go no further, but try again in 8-10 days. Whenever we find cells started in one of the strong colonies, every colony is examined. Queen cells are destroyed at all stages, whether advanced cells or only eggs, every ten days. If a cell contains a larva or an egg, it is smashed with the hive tool—very slight defacement will cause the bees to reject the cell.'

It is obvious that this method of going through all colonies every ten days, entails a lot of labour if stocks are, as they should be, full of bees and brood. Mr Manley claims that it is easier to detect queen cells than commonly supposed. He says he has only to lift the super up from the back to find them, for they are either at the bottom of the top chamber, or the top of the lower. This may be so, but the fact remains that if *any* cells are found, practically every comb must be searched to make sure all are destroyed, and since bees make a point of clustering thickly round queen cells, it is easy to overlook them unless the combs are shaken.

Dr Miller admitted that even persistent destruction of cells will not always stop swarming, but may even harden bees in their determination. In fact, in my view, this is an addition to the unsettlement already present and may aggravate it so that the swarm will come out before queens are mature. The process having been started, nature urges that it should be consummated. You cannot stop the formation of flowers on a plant by picking them off; in fact, you can often induce more flowering by so doing.

ARTIFICIAL SWARMING

Artificial swarming is an attempt to forestall the bees' natural irruption from the hive. It consists in the division of colonies which have reached the prosperous condition precedent to natural swarming and may be carried out by the Pagden, Demaree or Snelgrove plans, the object being, as when dealing with natural swarms, to separate queen from brood. In this way swarming can infallibly be prevented, or at any rate, postponed *sine die*, but unless the greatest care is taken, the result will be far from that desired. Instead of gathering nectar, the bees will try and restore the disturbed balance of the organization.

60

In my opinion, artificial swarming is best done as an adjunct to the cell-killing plan, colonies being divided *only when queen cells are in an advanced stage.* It is the greatest possible mistake to interfere with colonies which show no sign of imminent swarming. They are almost certain to be the best honey-gatherers and to divide their forces, even by the most approved methods, will cause disorganization and loss of crop. It is better to give such stocks more breeding space, either by the addition of a brood-chamber, or by removal of combs of sealed brood, giving these to weaker colonies and providing empty combs or frames of foundation. Even this should be done only to a very moderate extent, for every such change disturbs the rhythmic order of the hive.

Of course, if the bee-keeper is a novice with only one hive, he may quite reasonably prefer to increase it to two or more rather than try for a good crop of honey during the first season. This is the plan often followed by shrewd gardeners, who buy a single plant of some choice fruit and sacrifice a possible crop for the sake of propagating more plants from it. This is quite legitimate and justifies the division of a strong colony at the earliest moment, even when it shows no sign of swarming. It is not in fact the best way of making increase, which, in a fair season, can be done without sacrifice of the entire honey crop, but it has its advantages for the novice, so I will describe this crude and simple method. Fit an empty hive with four to six frames of foundation and stand it beside the stock to be divided. From this stock take out three combs with the adhering bees and put them in the middle of the new hive, with the frames of foundation on either side. Make sure there are eggs in at least one of the combs, but also be sure some are left in the old stock. The old stock is then removed to a new site not less than 6 ft away and the new one put in its place. Flying bees will return to the old stand, so the new hive will contain only a few young bees and brood, but practically all the foragers. So long as there are eggs in both parts, it does not matter which contains the queen, for either has the means to raise one.

These are the only *essentials* and there is little risk of failure during the swarming season, but several improvements can be made. It is better that the foraging colony should not have to raise a queen, so it is an advantage if she is in the new hive. When the division is being made, she should be found and only the comb she is on put in the new hive, the remaining space being filled with frames of foundation. If any queen cells are on this comb they should be destroyed, or the comb exchanged for one without them.

FINDING THE QUEEN

The novice often has great difficulty in finding the queen in a populous stock and if she is not found at the first inspection, the division should be made as suggested and the entrance of both hives watched carefully. In a little while, say ten to twenty minutes, it will be seen that there is much agitation in one of the hives, bees running up and down the front. This shows that the queen is in the other hive. If this is the one now on the old stand all is well and nothing more need be done. If she is on the removed part she must be found and transferred to the other, either by picking her off the comb, or, as most beginners will prefer, by exchanging the comb she is on with one of the other three.

Any of the methods described under swarm management, the Pagden, Demaree, Snelgrove or Shepherd plans, or suitable modifications of them, are equally applicable to artificial swarming but it is, as I have said, most important if a crop of honey is to be secured, not to carry out the work too soon. It does not follow, because a hive is 'boiling over' with bees, that it will soon swarm. I have known such stocks go right through the season without a sign of swarming, even though they had not too much room. Once the main honeyflow has started, they will build comb and store honey in every available space, rather than swarm. I have often seen the intermural space in a double-walled hive filled with honeycomb, and once I found it built beneath the floor. Division of such a colony is disastrous and will ruin the crop. It should be given plenty of room, piling more and more chambers on top. Never divide until queen cells have been started and it is better, even then, to wait till the first one is sealed over.

Transferring selected combs from an established colony to form a nucleus

RAISING A NEW COLONY

If it is desired to raise a new colony, the swarming season is the best time to do so. Two combs, including one with a sealed queen cell, are put into a new hive and the bees from another comb or two shaken in with them. This will be the nucleus of a new stock which can be built up gradually during the summer by judicious feeding. All queen cells in the old stock must be destroyed and frames of foundation given in place of those removed.

SWARM CATCHERS

A method which sounds well in theory is to use swarm catchers. These devices vary in pattern, but their principle is the same. In front of the hive entrance, an auxiliary chamber is fixed; it may be small, or large enough to contain two or more standard combs. The exits from it are covered with queen excluder, so that workers can come and go, but a queen cannot. She remains in this ante-chamber and the swarm clusters about it. It should be removed and hived as soon as possible.

In order that swarm catchers shall be entirely effective, every hive must be fitted with them—a considerable expense—or hives must be examined first for signs of swarming. In the latter case there is little advantage, for other steps can be taken to cope with the situation. For this reason, swarm catchers have never become popular. In addition, excluder hinders foragers a good deal and interferes with the hive ventilation—an important matter at this time of year.

CLIPPING THE QUEEN'S WINGS

Another device intended to prevent loss of swarms is clipping the queen's wings, so that if a swarm issues it remains close to the hive because the queen cannot fly. Those who practise clipping, operate while the queen is in the nucleus, as soon as the production of worker brood shows that she has been duly mated. Fine sharp scissors are used and the large wing on one side is cut, about the middle. To do the work safely, the queen is picked up by the thorax—never by the abdomen. One blade of the scissors is slipped under the wings, care being taken not to include a leg. Another plan is to use a little wooden fork, across which a piece of thread is stretched. This is laid over the queen's thorax as she walks along the comb. Before clipping a queen for the first time, you should practise on a few drones. There is always some danger of bees superseding a clipped queen—perhaps because of the mutilation.

MARKING THE QUEEN

Some bee-keepers mark their queens by putting a spot of bright cellulose paint on the thorax, or using a device called the Eckhardt marker, by which a small piece of brightly coloured tinfoil is placed on the thorax. This makes a queen easier to find and if a different colour is used each year, her age is known.

DETECTING THE DEPARTURE OF A SWARM

The most experienced and successful bee-keepers may have prime swarms now and then, but the occurrence of second swarms or casts, is a sign of ignorance or neglect. Unless the apiary is outside one's back door, it is always possible for a swarm to leave without being seen. In about ten days' time another smaller swarm may be sent out by the same stock. This will have a virgin queen and because of her youth and vigour, the swarm will often fly straight away without pausing in the vicinity, as a prime swarm almost invariably does. In this way a stock is deprived of its foraging bees and gets little surplus honey. Swarm catchers and clipped queens are designed to avoid this, but other precautions can be taken to prevent this grievous loss.

Every day during the swarming season a good look round the apiary should be taken towards evening. Once a swarm has clustered, it keeps very still until ready to move off and will seldom catch the eye unless you are looking for it. Hedges and bushes and other likely places should be carefully inspected, especially those which have been the temporary resting-place of former swarms. A swarmed stock can often be identified by the obvious decrease in the numbers of bees coming and going. If young bees are seen crawling round a hive it may be that they have come out with a swarm and, not having flown before, were left behind.

If there are supers on the hive, it is generally possible to discover whether a swarm has left by peeping in the top feed-hole late in the evening. During the daytime there will not be much difference, but towards night bees go down to cover the brood below, sometimes leaving the supers empty.

To be able to read these signs, the bee-keeper must keep an intelligent eye open every day during the active season, so that he can compare the appearance of stocks one day with what they looked like the day before.

If there are several stocks and a swarm is found clustered outside, it will not be known which hive it came from and none of the swarm control methods can be applied until this has been ascertained. Fortunately, there is a certain method of finding out. After the swarm has been collected, it should be put in a shady place until evening, when nearly all bees have ceased flying. It should then be moved to a new place at least 20 ft from the hives and a handful of bees shaken out on to a board or sack. These bees are then sprinkled with fine flour and the swarm either moved somewhere else, or covered up with sacks, so that they cannot locate it. After a time the floured bees fly off, going first to the place where the swarm pitched; they circle and search there for a time and, not finding the swarm, return to the hive from which they came. The flour enables them to be easily seen and the desired treatment can be undertaken next day.

If, for any reason, a hive is left with ripening queen cells, a cast is likely to be thrown within ten days. If this is captured and the hive from which it came certainly known, it should be kept till dusk. The combs should be inspected and any remaining queen cells cut out. In the evening the front of the hive should be tilted up and the cast thrown on the floor. A couple of wooden blocks can be used to prop the front up till the bees have climbed to the combs. Any other queens will be killed and thrown out by morning.

Casts are often useful to build up a stock, for if fed and later given sealed brood from other stocks, they form good colonies for next season, since they are headed by young queens.

5
The Honey Harvest

It is true to say that in Britain the active season of the bees' year consists of three parts. During spring, from March to May, when the tide of blossom is swelling, the colonies grow at an increasingly rapid rate. It is not at all unusual for a stock which, at the beginning of March was occupying not more than three or four of its combs, to have the whole brood-chamber full by the middle of May.

The second phase is comparatively short, though it varies a good deal. It lasts from the time May-bloom begins to fall, till the white clover comes out. This is generally known as the 'June gap' and the first three weeks in that month is the usual time for it. It may be shortened in some districts because of special crops, such as raspberries or sainfoin, or there may be a handy field of beans to give a welcome and heavy income for a few days. Whether long or short, this gap coincides with the time when bees are near their maximum strength. There are not only vast numbers of adult workers and drones to feed, but a very big nursery to attend to: every scrap of income is therefore needed for maintenance and the accumulation of stores must be checked. If the weather is unfavourable—June is sometimes a wet, windy or even cold month—the stores will be drawn on heavily and supers full of honey at the beginning may be empty by the end of the month.

LEAVE THE EARLY CROP

For this reason, the experienced bee-man never takes away the early honey— a mistake the beginner often makes—which would check the full development of the colony. As I have intimated, it is in this gap of comparative dearth that swarming is most likely to take place. Once it has passed, the chance of it is much reduced, though it does sometimes happen at the very height of the main flow.

The whole art of bee-keeping is to produce the strongest possible force of foraging bees just when there is the largest amount of nectar to be gathered. In some places—such as the fruit orchards of Kent and Worcestershire—this is the month of April, but it is impossible to bring a colony up to full strength by then, for there is no means of short-circuiting the six weeks needed to bring bees from the egg to the foraging stage because, even if bees are artificially fed, the temperature in early spring is too low to induce the queen to lay sufficiently fast.

TO SECURE SURPLUS HONEY

There is a method by which the owner of several hives may, if he wishes, take advantage of the fruit blossom to secure surplus honey. Two hives are brought close together—by stages of a yard a day, if they are not already near enough to do so in one hop. Then, at the moment when the bloom is at its best and the bees are going all out to get as much as they can, one hive is moved right away and the other placed midway between both stands. Additional storage room is given to the latter and the double force of foragers will ensure a considerable increase in the amount stored. It is essential to do this only on a fine honey-gathering day, otherwise returning bees not belonging to the doubled stock may be prevented from entering. Some bee-men add also a number of combs of sealed brood from the removed stock, to provide a good supply of 'house bees' and thus release more foragers.

This is not recommended for general use and in any district where the main flow is from clover, it would, in the long run, be bad practice for two reasons: the hive denuded of adult bees would not build up so effectively for the main flow, while the bees added to the doubled hive would be worn out by their labours and useless for the clover. Consequently, instead of having two strong stocks for the main flow, there would be one strong and one only moderate. The stocks which gather most honey are those with a great preponderance of foragers—in simple words, only surplus bees produce surplus honey. I repeat that, in the long run, it is sound policy to leave in the hive all honey gathered before the June gap, so that its strength may be maintained and increased to the maximum by the time the main flow arrives.

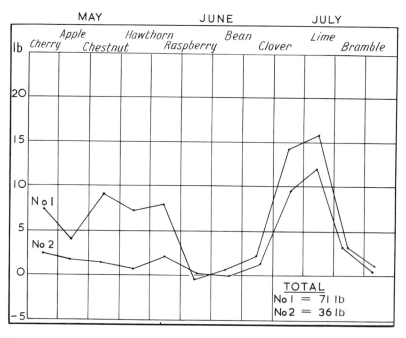

Chart of honey season, showing weekly net gains of two hives—one weak and the other strong

THE MAIN FLOW

The third and final phase is the main honey-flow which, in all but moorland districts, begins with the clover and ends with basswood trees. Other plants often come into the picture, such as sainfoin already mentioned and charlock, that bright yellow weed which sometimes smothers the corn.

The duration of the main flow is immensely variable. Wild white clover, by far the most valuable of the tribe, may begin blooming as early as the end of

66

May, or not until a month later. It continues flowering down to October, but rarely in sufficient quantity to provide surplus after July. For this plant, ideal conditions are a light soil with plenty of lime and the chalk hills in various parts of Britain have an enviable reputation for clover honey. In districts with heavy clay soil, though the main flow lasts longer, it is less reliable, for such soils do not warm up quickly and only in really hot summers does the clover respond fully. On the other hand, in summers of continued drought clover is soon over on light soil and the flow may not continue more than two or three weeks. A cool showery period at the beginning of July, providing it does not last too long, will often prolong the blossom and consequently the honey-flow. During the early days of July, basswood blooms freely and yields a good crop; in suburban districts, it is often the main source.

No precise time can be laid down for any of these things, nor are any two districts alike in their bee flora. Once having grasped the general principles which govern the honey-flow, the bee-keeper must use his own intelligence and observation of the potentialities of his district.

EQUIPPING THE HONEY HOUSE

More beginners get into trouble about removing the honey than from any other cause. They suppose, not unreasonably, that the bees will resent the operation and they are apt to put it off as long as possible. This is the very worst thing they can do. It may seem to the uninitiated that while bees are flying in large numbers they are most likely to be troublesome and that it is better to wait until a time when they are quieter. In fact the reverse is true, for while there is active foraging, there are fewer bees at home during day-time and these are comparatively young ones, much less likely to sting than the old hands. If all is quiet outside, the hive is packed full of bees and immediately the hive is opened, they rush out to repel marauders.

Furthermore, as I have explained, the secret of pacifying bees is to fill them up with honey by frightening them with smoke or other suitable intimidant. They cannot fill up unless there is unsealed honey in the hive and every day after the honey-flow is over, more and more honey is sealed and less is available for immediate refreshment. If the honey can be removed before the flow is quite over, it may be done without any trouble worth speaking of. The longer the work is postponed, the more the bees resent it.

A SUITABLE WORKSHOP

Before I go into details of the best methods and times to remove the honey I will describe the equipment needed, which must be all in readiness before the work begins. Bee-keepers with only one or two hives may be able to manage without a specially fitted room or hut for the indoor part of the work, but they will always find it very inconvenient. Honey is adhesive stuff and if dealt with in the ordinary kitchen it will not be difficult to mess up, not only the kitchen, but every other room in the house, for honey on the hands is easily transferred to door-handles. If any gets on the floor, which is not easy to avoid altogether, it may before long be on the drawing-room carpet and the bee-keeper will have to flee from the wrath to come. Moreover, in any well-run household the kitchen is sacred to the housewife and she may reasonably resent being turned out of it for an indefinite period. Yet another snag is that bees—and wasps—have an unerring nose for honey and will soon invade any enclosure not specially designed to keep them out. On these grounds, as well as that of efficiency, it is desirable to have a place solely devoted to apiary work and preferably quite outside the dwelling house.

Its size is not important, because its use can be adjusted to the space available. If it is to serve as a store for supers, spare hives, etc., it must naturally be larger than one used only for the work of extracting honey, fitting up frames and so on. In the latter event, one can manage in a space 6 ft × 6 ft, but I would say 10 ft is more satisfactory. If the place is not bee-proof, it must be made so, holes being stopped with cement or paper pulp rammed in. A badly fitting door should have a flange round it, while any large space you do not wish to close up entirely, for the sake of ventilation, should have wire cloth or glass so fitted that bees can get out, but not in. There are several ways of doing this, but the best is to fit a piece of glass half an inch shorter than the opening, so that the space is at the bottom. Half an inch *outside* this, another strip of glass, about 2 in deep is fixed. Any bees carried into the room will soon make for the light. They climb up the glass, but when they reach the top, drop to the bottom and climb up again. This time they walk up the narrow strip and so out: bees very rarely try and walk downwards, so they do not re-enter.

EQUIPMENT

Two important items of furniture are required, besides the actual machinery for extracting. The first is a work table, which should be firm and solid. For ordinary purposes, a wooden top is best, but for extracting, a marble or enamelled top is better. One can get a loose marble slab or sheet of enamelled iron, or failing these, a sheet of American cloth can be used. This should be fastened to blind rollers and rolled up when not in use, so that it remains quite smooth.

The other item is a small bee-proof cupboard in which supers of honey can be deposited, so that if the outer door is left open, they will be quite safe. In a small hut, such a cupboard can be contrived under the table.

The fewer gadgets you are encumbered with during extracting the better, but the following are essential tools.

The honey extractor is a cylindrical tank, provided with a set of cages attached to a spindle, which revolves in the middle of the tank, the power being transmitted by chain or cog gearing. Small sizes are operated by hand, but larger ones can be fitted for mechanical power. There are now three different types, the oldest and still the most widely used, being the 'tangential' type, in which the cages are parallel with the sides of the tank. These are made to take from two shallow frames (the smallest size) to four standard or eight shallow frames.

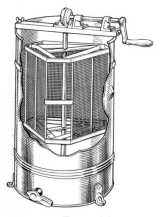

Tangential extractor

Another form, which has several advantages over the tangential, is the 'radial extractor', in which the cages are so arranged that the frames radiate outwards from the spindle. This type can be obtained to hold from eight to thirty-two frames, without much increase of size over the tangential form. When using the latter, only one side of the comb can be extracted at once and it is necessary to lift all combs out and turn them round to do the other side. The radial extracts both sides at once.

68

The third type is the 'parallel radial' machine in which the spindle is horizontal instead of vertical and the combs turn over and over instead of round and round. It is claimed that this machine extracts with less risk of comb breakage.

In either case the basic principle is that rapid revolution of the combs causes the honey to fly outwards against the tank wall and run down to the bottom, where it can be drawn off through a tap.

If the extractor is a small one, it can stand on the floor, but in this position the tap is too low down to allow a receptacle for honey to be put under it and it must afterwards be lifted onto a stand for this purpose. For a larger machine it is essential to have a stand raised at least a foot from the ground. It must be heavy, as the machine is liable to rock at high speed. Where the floor is of cement, another plan is to dig out a pit to hold a honey tin or pail. One advantage of the parallel radial is that its outlet is well above the ground.

Formerly called a 'ripener', the storage tank is a tall metal vessel, holding from 56 lb upwards, into which the honey from the extractor is transferred. It has a large container at the top with a perforated bottom to strain the honey and retain the coarser particles of wax. This also is fitted with a tap at the bottom. The advantage of this vessel is that straining goes on without attention and in perfect safety from bees or wasps. A small bee-keeper can manage without one of these articles, if he has a sufficient supply of large honey tins, holding 14 lb or 28 lb, but he will then have to strain his honey in the open by hanging a cone-shaped strainer on the extractor tap.

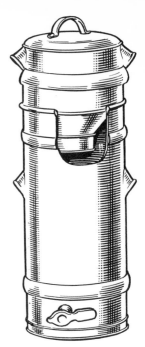

Honey storage tank

Before combs can be emptied of honey, the cappings must be removed. For this purpose a long sharp knife is required. A good carver is little inferior to the special knives sold and a saw-bladed bread knife is also an excellent tool for the purpose. Two such knives should be used and they must be kept hot by standing them in a vessel of hot water. A rather more costly knife can be used without heating, and knives heated by steam or electricity are also obtainable.

Bingham knife for uncapping

Uncapping has to be done over a vessel to catch the honey and wax removed. A rather elaborate, but useful, uncapping vessel is the Pratley tray, which is fitted with a water jacket, kept hot by a lamp beneath it. This melts both honey and wax as it comes from the knife and they run off into a vessel placed below, the wax rising and forming a cake on top of the honey.

Pratley uncapping tray

If expense is an obstacle, one can manage very well with one or two large meat tins, fitted with draining grids. As these are filled, the contents are turned into the storage tank top. A paper hanger's scraper is, I find, the best tool to scrape honey from tins, or for lifting masses of sticky cappings.

REMOVING THE HONEY

One of the most useful inventions connected with bee-keeping is the Porter Escape. This is a metal tunnel fitted with a pair of light springs, which meet at one end and afford passage one way only. The bees enter the open end of the tunnel and press the springs apart as they pass through. The springs close behind them, so there is no return. This appliance is fitted into a board large enough to cover the brood-chamber. It is an advantage to have two or more such escapes in the board, so that more bees can pass at the same time. The apparatus is used solely to clear honey supers of bees before removal, and is therefore known also as a 'super-clearer'.

Uncapping with an electric knife

When the time has come to remove the honey, it is best to choose a fine day when bees are busy, but it can be done at any time after the honey is sealed over. Unsealed honey is not properly ripe and if extracted at this stage will be thin and very liable to ferment in store. A few open cells in a well-capped comb do not matter, but the bulk of them should be fully sealed.

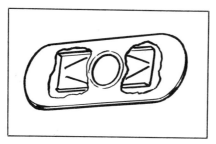
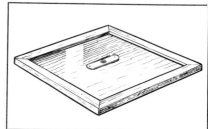

Right: Diagram of Porter Escape showing one-way mechanism
Far right: Super-clearing board with Porter Escape inserted into the feed-hole

OPENING UP

Having removed the roof of the hive and turned it upside down, as was done when adding supers, stand the super-clearer on it, making quite sure it is right way up, so that bees can only pass down from above. Insert the hive tool under a corner of the super, lever it up and put in a piece of wood to support it while the other corner is lifted. It is best to choose the side which holds the frame ends. Puff in a little smoke and gently raise the super until you can see the frame ends. If any of these are sticking up, they must be pressed down and no attempt should be made to lift the super until all are quite free. Holding the super firmly—it may weigh 30 lb, so it will not be a light task—give it a slight twist to free it finally from the brood-chamber and stand it on the clearer. Then lift board and super together and put them back on the hive. When you have made quite sure that there are no gaps round the super through which bees can pass, put the roof on and leave the hive alone for some hours. The time taken to clear a super varies, but if put on one day, it will be clear the same time next day. It can then be removed—leaving the board in place for the time—and taken into the honey house.

If there are two or more supers on the hive they can all be cleared at once by putting the super under the lowest, but if they are both full, they will weigh little short of 60 lb, so that it is a task for two persons or a very strong one.

This is the safest, most foolproof method of removing honey, but it has serious drawbacks. It is always desirable to get the honey extraction done in one operation and it is much easier if done in the middle of a warm day, for honey flows more readily then. If the clearer is put on in fine weather, it may be much cooler the next day when the super is removed. Moreover, if there are two or more supers on a hive and you cannot lift all at once, only one can be removed a day. If there are several hives, an escape board must be provided for each, or the operation will be still more protracted.

DIRECT REMOVAL

For these reasons, many experienced and large-scale bee-keepers dispense with escape boards. For many years I have used them only for sections, which cannot be dealt with in the same way as extracting combs.

70

I extract on the first warm fine day after the honey is sealed, preferably before the flow is quite over. There is not usually much danger from robbing for a week or so after the end of the flow, but it is best to be on the safe side, for any honey which comes in afterwards is valuable winter food for the bees and will reduce the amount of syrup feeding required.

I provide myself with some empty super boxes and a few clean burlap bags. I make sure the smoker is well alight and with plenty of fuel to last an hour at least. The hive tool and a bunch of grass for use as a brush are the only other requirements. Beside the hive to be dealt with first, I spread a bag on the ground and on this put an empty super box. I remove roof and inner cover, which I put upside down with the bees on it in front of the hive. Having puffed smoke into the super, I loosen all the combs with the hive tool so that they come out easily. One by one I take out the combs and shake them over the board in front of the hive. Any bees left on the comb I brush off with the other, but in practice, so few bees remain on sealed comb that they are cleared with surprising speed. As each comb is cleared, I put it in the empty super and when this is full, cover with another bag. I then remove the empty super from the hive. The second or it may be third super, is dealt with in the same way, an empty super being put on the full one to receive the combs. Finally, the inner cover is replaced on the brood-chamber.

Each hive is dealt with in the same way, so that at the end there is a series of supers beside each hive, all safely covered with sacks. I have never found robbing occur during the process, but, as I say, I generally do it before the honey-flow is quite over.

Two other methods of removing bees from supers must be mentioned: repellent chemicals and forced air currents.

There are a number of volatile chemicals which bees do not seem to like and from which they retreat. Two of these, proprionic anhydride and benzaldehide, have been used with some success for clearing supers. The method of application is somewhat similar in both cases and consists in having some kind of absorbent pad in a hive-sized frame on which the chemical is sprinkled. The supers are then lightly smoked from above and the 'fume chamber' placed on top of the top super. The bees should move down about the depth of one super. The difficulty is that sometimes the bees are not so co-operative and refuse to go down; they will certainly not do so if there are open cells or if the weather is cold. There is also the risk that the powerful odour of the repellent may cling to the supers and be faintly detected in the extracted honey.

The second method is to have a fan driven by an electric motor—rather like a vacuum cleaner in reverse—which will physically blow the bees out. This method is used by some American honey producers but it is clearly only practical in large-scale honey farming and the effect on the temper of the bees is problematical.

A COMB CARRIER

Quite recently, I have improved on the first method. Formerly, I used to collect all the supers in a barrow to convey them to the honey house, but a barrow is rather an awkward thing for the purpose. One day I had a brain-wave. We had a deep old baby carriage, long standing idle, and it

71

struck me that if the inner fittings were taken out, it would hold combs. Measurements confirmed this, so it did not take long to convert the vehicle into a comb-collector. Being an old-fashioned deep pram, it held two rows of combs—forty shallows in all. Ledges were fitted inside to take the lugs, and a couple of sacks served to cover the whole thing. Instead of piling the supers on the ground, I put each cleared comb straight in the pram, moving it from hive to hive till it was full and ready to be pushed into the honey house, the combs transferred to empty racks and the pram pushed out for another load.

By the time the hives were cleared, and the combs safe in the house, the bees in front of the hives had climbed in, so I went round, removed the sacks and replaced the cover boards and roofs.

Not only is this plan speedy, but it cuts out the hardest part of the work— lifting the heavy supers. This has to be done twice when the clearing board is used.

The board *must* be used for sections, not only because they cannot be shaken in the same way as combs, but also because you must not take the smallest risk of robbers, who would tear at the cappings and spoil their appearance, whereas a few damaged cappings are of no consequence with extracting combs. Sections also must be completely sealed, so they usually have to be left till the honey-flow is over. In using the clearer it is most important to see that bees cannot get in or out other than through the escape.

EXTRACTING HONEY

Having carried the honeycombs safely into the house, begin the work of extracting at once. Be sure the machine is quite clean and runs smoothly. It should be so placed that the uncapped combs can be put into it without honey dripping on the floor. If the table stands, as it should, in front of the window, the extractor can be either to left or right and close up to it. The right is best, as the operating handle will not be over the table. The side near the table should have half its cover removed—the other half can be left on. On the table, stand the uncapping tray and just behind it the jug of hot water with the two uncapping knives. Have the scraper handy to scoop up any honey which may fall on the table and be likely to run on the floor. It is a good idea to have a bucket of water and some clean cloths in case your hands get too sticky.

If all the combs seem much alike in colour when held to the light there is no need to pick and choose, but if there are any specially light ones, they may be extracted first, if one wishes to have—say for show purposes—a few jars of light honey. When these have been extracted, the honey may be run off and kept separate.

Uncapping is not difficult when the knack has been acquired. Holding the frame by one lug, rest the other on the support provided in the tray, taking care that it does not slip. It should be tilted over a little at the top on the side being cut. The hot knife is placed flat on the frame below the comb and drawn upwards, so as to slice the capping off in one pice. The tilt will cause it to fall away from the frame and drop into the tin. The comb is then turned round and the other side cut in the same way. If parts of the comb do not project beyond the wood they will have to be cut with the end of the knife, after the main slice has fallen. Some bee-keepers prefer to use an uncapping fork. This useful instrument has some twenty sharp tines about 2 in long, often slightly

72

cranked. The points of the tines are inserted in the comb just below the cappings which can then be removed with great facility and a minimum of mess.

Still holding it by the lug, put the comb into the cage of the extractor. If this takes two shallows on a side, see that the first is well up to the end, or the second may not go in. Also make sure it is right down to the bottom, for in some machines a frame that sticks up slightly is likely to have its lug broken off when the cage is revolved; sometimes the metal ends get in the way and if they are at all loose it is better to remove them before uncapping. When the full number of combs the machine takes are in, the handle should be turned, gently at first and then faster till the honey can be heard pattering against the side of the machine. In a radial extractor, continue until it is judged that the combs are empty. In tangential machines, the combs must be turned round. Indeed, if the combs are new, it is wise to run the machine gently and get part only of the honey from the first side, turning the comb to do the other and then turning again to finish the first side. Old tough-wired combs will stand up better, but too much pressure will break the middle out of new combs. To reverse the combs when each is in a separate cage, they must be lifted up, turned round and dropped in again. When the cage is, as in some models, continuous all round, only the first need be lifted out, the others being turned *in situ* and the first one put back at the end. The emptied combs are put back in the racks and other combs extracted until all have been done.

Meanwhile, the honey in the extractor must be drawn off as it accumulates and comes up to the level of the cages. It can be drawn into large tins or a pail and poured into the straining tank. If one has no tank, it must remain in the tins and should be strained as it comes out of the machine. Great caution is needed in doing this, for honey runs slowly through the strainer and, if it is allowed to come out too fast, may run over the top or overfill the tin, making an unholy mess on the floor. It is a wise precaution to stand the tin in a good-sized meat dish. Honey does not gurgle like water, but runs quite silently and, if one is doing something else, it is easy to forget it. A proper straining tank is, for this reason alone, a worthwhile investment, as the extractor can be emptied quickly and there is no need to leave it running. Also everything can be put into the tank, including cappings and broken comb.

Cleaning up is a job which should not be put off after the honey has been cleared from extractor and tins. If there is more than can be done in one day, there is no harm in leaving the things as they are for a day or so, but when all has been extracted the apparatus should be cleaned up at once. It is a simple matter to let the bees do it by leaving the door of the honey house open: the moment a tin smeared with honey is put within reach, bees and wasps will arrive and lick it clean in a matter of minutes. But there is a snag in this. If one were sure that all one's bees were healthy, it would be quite safe to do this, but there is nowadays rarely such assurance, and the possibility of attracting strange bees, which may come from a diseased hive, makes it rather dangerous. There is no harm in giving back to a hive its own honey, but since all has been unavoidably mixed, any from a diseased hive will be distributed throughout the apiary. For this reason it is as well, when removing supers, to chalk on their sides the number of the hive and return the supers to the hives they came from.

To clean out the extractor, close the valve and pour in a gallon or so of cold, or not more than tepid, water. Swill this well round until all honey and wax has been washed off. Run the liquor into a tin and use it to rinse out the un-capping tin. This liquor can, if one wishes, be saved to add to the washing of the cappings later on, to make mead (honey wine) or honey vinegar.

Hot water can then be used to clean out the extractor, which should be dried thoroughly and stored with a cloth of some kind over it. It is an expensive piece of apparatus and will last a lifetime if proper care is taken of it, but it will soon be ruined, if left with honey adhering to it.

One thing more remains to be done—the return of wet combs to the hives from which they came. This should not be done till evening, otherwise the smell of honey will attract robbers. The roof is removed and the hole in the cover is exposed sufficiently to let the bees come up. The super is put on, covered with a sack and the roof placed over it. In a few days the bees will clean and even repair the combs so that they can be stored away for use another year. They should be wrapped up in newspaper immediately, so that wax moth cannot reach them, for they are a most valuable asset. It is said that combs stored wet are not attacked by wax moth, but I have never tried it myself as I much prefer to have the combs clean for use when needed.

The cappings in the strainer tank will continue to part with honey for a few days if it is stirred up from time to time, but when honey no longer runs out the cappings should be turned into a large vessel with tepid water and allowed to stand for twenty-four hours, being stirred round occasionally to get as much honey out as possible. The liquor can be drained off and put with the other washing water to make mead. The wax is dried thoroughly and either melted down at once, or stored in a place safe from wax moth till it is convenient to deal with it.

Sections should be left till they are fully sealed before removal. In a really good season this does not take long, but in moderate or poor years they may have to wait a week or so after the flow is over and even then some will not be finished. They should not be left on after they are sealed, or they will be stained by the addition of propolis and added wax.

After removing them by the aid of the clearing board, they must be taken from the rack carefully, as they are easily damaged. The best plan is to use pieces of wood about an inch square and long enough to go along a row of sections, inside the rack. After removing the spring and board which hold the sections tight, put the strips of wood under the rows, and press the edges of the rack down. This forces the sections up above the rack and they can be easily removed. Propolis and wax on the wood should be scraped off with a blunt knife and the sections packed temporarily in a tin. Standard biscuit tins hold exactly sixteen. The final preparation for sale will be described later.

6
Autumn
Work

When the harvest has been removed, the bees should be examined without undue delay, so that anything necessary may be done to get them into good condition for winter.

Normally there will not be very much brood at this time, as the concentration on honey-gathering causes the queen to be pushed into the background and sometimes she stops laying altogether for a while. But there is no rule about it and you may sometimes find the brood-chamber full of brood. Very often, after the supers have been removed, there is neither honey nor brood in the hive. If this is found to be the case, the first thing is to make sure there is a queen. If one is not found, steps must be taken to provide one, if the colony is otherwise strong. The most usual way to make good loss of queens at this time is to unite the stock to a nucleus, raised by the methods explained in the section on queen rearing. Otherwise a queen must be purchased. At this time of year they may be had relatively cheaply.

If the stock has a queen, she must be induced to begin laying again. Remove one of the empty combs—this is one occasion when damaged, irregular combs, or those having too many drone cells, can be culled. This makes room for a frame of foundation, which should be put in the centre. Queens always prefer to lay in a new comb and this frame should be filled with eggs as fast as the comb is built out. It is essential to feed the bees, so a jar or can of syrup is put over the feed-hole, which will be immediately above the new frame. In a week's time the stock should be inspected to make sure that laying has begun. Steady feeding for a month or five weeks will generally ensure the raising of a good batch of young bees, the first essential to successful wintering of the colony.

THE BROOD-CHAMBER

It may be, especially after a particularly good honey season, that the brood-chamber is full of honey, so that there is no room for the queen to lay; if this happens the same treatment should be given, one of the sealed combs being removed and a frame of foundation put in the middle. In this case it will not be necessary to feed, since the bees will make use of the stored honey in addition to that which is coming in. The removed comb of food can be given back later on to make up for what has been used. Now and again, it will be found that the queen of a prolific race, like Italians, has filled the chamber with brood, while all the honey has gone into the supers. Providing the weather is reasonably good during August, nothing need be done for the moment, for the bees will be able to get sufficient for daily maintenance and the brood will steadily emerge and add to the foraging force. Special care must be taken to give such a stock an ample supply of syrup before winter and, if an abundant crop has been taken, it is provident to reserve a full super of honey to put over the brood-chamber for winter.

It is a saving of time and trouble if, at the time the honey is taken, the brood-chamber is examined and a note made of its condition. This is readily done while the bees are still massed outside the front of the hive. You will then know, without further inspection, what treatment is needed and those which are very short of food in the brood-chamber can be given back a full super.

My own practice is to use a card for each colony and to note on it: how many combs are covered by bees, how many filled with brood and the approximate amount of food present at every examination. By doing this I know exactly how much food it is necessary to supply.

FEEDING BEES

There is a saying that 'bees work for nothing and board themselves' and it is true that they do not, like other domestic stock, need to be fed regularly and constantly. Because of this, many beginners fail to realize that judicious feeding is one of the most important aids to profitable bee-keeping. Even in bad seasons, bees rarely fail to get enough to carry them through the winter, but as the bee-keeper wants honey for himself, they must either gather very much more than they need for winter, or something must be given them in lieu. It has been conclusively proved that the more food bees have in store in the autumn, the better progress they make in spring and consequently the more surplus they gather the next season.

From October to March food consumption in the hive is small and steady. It makes very little difference whether the winter be severe or mild, during those six months about 10 to 12 lb of food is consumed. This I have proved by many years' tests with hives on scales. After March, food consumption increases rapidly, owing to accelerated breeding, and although in fair weather there is no lack of nectar in the fields, the foraging strength is not yet great enough to take full advantage of it. I have rarely found a colony increase in weight before the end of April. As a rule, there is a continuous fall, although some of the weight lost is balanced by the increase of brood.

If spring conditions are bad, restricting income while the queen is laying rapidly, the loss during April is great and a really bad patch of weather coming then may mean the death of the stock if it has not a large reserve. That is why all authorities agree that, to be quite sure bees have enough to carry them safely through to summer, at least 30 lb of food should be in the hive at the end of September.

SUGAR v. HONEY

Undoubtedly the best food for bees is honey and some bee-keepers set their faces steadily against artificial feeding. I know of no substantial evidence that commercial sugar is harmful, although it has been shown that bees fed entirely on sugar do not raise brood so satisfactorily as they do on honey. The ideal would be never to take more honey than will leave the colony the 30 lb required for winter and spring, but in some seasons this would mean nothing for the bee-keeper, and the difference in money value between honey and sugar is so great that few can resist the temptation to substitute at least a part of the honey crop by sugar.

AUTUMN FEEDING

In practice, autumn feeding is a regular necessity, especially because most colonies reserve the brood-box for breeding and put almost all surplus in the supers. After these have been removed there is nothing left in the hive. Only poor and backward colonies are likely to have sufficient in the brood-chamber and, unless requeened, such colonies are not worth saving.

If the honey crop has been good, I strongly recommend that, in the absence of abundant stores in the brood-box, a shallow super full of honey be left on the hive. Food given to bees is never wasted and will be returned with interest the following season.

Except in heather districts, feeding should begin in September and be completed as rapidly as possible, for the cooler the weather the more reluctant the bees are to take syrup. As soon as the surplus has been removed and steps taken to ensure a good winter population of young bees, an estimate should be made of the amount of food needed by each colony. A standard comb solid with honey weighs about 5 lb, so there must be the equivalent of six full combs. Usually only one or two on the outside are full, but others may have more or less on top, so that it may appear that they have abundance when in fact there may be very little. The simplest and surest test is to lift the hive from its stand. If this can be done easily, there is obviously little food, but if it is really heavy, all is well.

A brood-chamber chock full of honey is not in a good wintering state, because it leaves no room for close clustering. Allowing, as I said, six full combs for food, that leaves only four (five, in the Modified Dadant) with empty cells, which is usually adequate, but if there is a greater weight of food in the brood-chamber, one of the outside combs can be removed and an empty one put in the middle. On all grounds, it is far better to leave the honey and supply clustering space by putting a shallow rack of empty combs, or even an empty rack, under the brood-chamber. This provides the required room for clustering and greatly improves the ventilation.

METHODS OF FEEDING

Bees may be fed in many ways, but the simplest and best plan is to feed through a hole in the hive's top. This is provided in wooden cover boards, but if cloth quilts are used, a flap about 3 in square is made in the middle by cutting through on three sides, so that the flaps can be turned back for feeding. The rapid feeders supplied by appliance makers are round vessels holding about 5 lb of syrup. There are some patent feeders made to hold up to 20 lb, but they are rather expensive and I can see no advantage in them over the simple lever lid can pierced with holes used by commercial bee-keepers. Any convenient-sized can will do, but if it holds about 10 lb it will need filling at most only three times. Some make the holes in the lid, turning the can upside down over the feed-hole, but I prefer to make them in the bottom of the can, so that it can remain in place until feeding is finished. The holes must not be large enough

Round feeder

to let the syrup run through quickly: from ten to twenty, made with a needle, will do. It will be necessary to put an empty super on to hold the feeder, so the can should be wider than it is high. The local pharmacist often has cans of all sizes to throw away, but it is essential to make sure they have not contained anything harmful. I used to get cans which had contained somebody's patent flour and they lasted for years. Most of the special feeders supplied have one big drawback: the bees have to come right out of the brood-chamber into the feeder, which they are naturally reluctant to do in cool weather, whereas the simple tin delivers the syrup right at the top of the cluster.

The bottle feeder illustrated in some catalogues is not worth buying, as it is never necessary to restrict the amount bees can take. If they are in a mood for taking syrup they will put it in their combs and use it as required and there is no point at all in doling it out through one or two little holes. To feed a small colony, like a nucleus, with a pound or so of syrup, there is nothing better than a glass jar or other container covered over with muslin.

When feeding bees, the following rules should be adhered to:

1 Feed only after sunset.
2 See that feeders are not accessible from outside.
3 Make sure the holes are not large enough to permit honey to run down and out of the hive.
4 Contract the entrances of all but strong stocks.
5 Take great care not to spill syrup about the apiary.

SYRUP, CANDY AND POLLEN

Syrup for winter feeding is made from 10 lb refined white sugar and 5 pints of water. Bring the water to the boil in a large pan, put in the sugar and stir until dissolved. Syrup of this strength can be stored safely in tins and used as required, so it can all be made at one time.

The time taken by bees to store the syrup in their combs depends on the strength of the stock and the temperature, but in early September a strong colony should take it in a week or ten days. To feed liberally and rapidly is the rule in autumn, for the sooner bees are brought into a quiet restful state, in a well-stored hive, the better. Every writer on bee-keeping stresses the importance of this, but tons of bee candy continue to be sold, because the advice is not followed, and in mid-winter the neglectful novice begins worrying whether his bees have enough food. There is no valid excuse for this, but it happens and one must make allowances for it. Bee candy can be bought or made according to this recipe:

Put 6 lb refined sugar into a pint of boiling water and add a teaspoonful of cream of tartar. Boil up, stirring constantly until the sugar is melted. Simmer for ten minutes and then allow to cool to about 120 °F. Stir the mixture until it thickens and pour into suitable boxes or shallow pots.

On a mild day—it is useless and foolish to do it in icy weather—uncover the feed-hole and slide the box of candy into place.

From what I have said, it is clear that spring feeding is not necessary if a full supply has been given in autumn, but sometimes a little judicious stimulation will urge a colony to increase breeding. Syrup for this purpose is made rather thinner than for autumn feeding. Its exact strength does not matter, but

Mixing syrup using an electric drill

7 pints of water to 10 lb sugar is usually recommended. This strength is also right for nucleus stocks or colonies engaged in queen rearing. Sometimes, also, a period of poor weather in summer may be made profitable by giving bees frames of foundation and feeding to induce them to build them out. If fed liberally, a stock will draw such combs out evenly and fill them with syrup. When sealed, they can be kept to supply autumn deficiencies, but care must be taken not to get them mixed with real honeycombs.

Bees require pollen as well as honey and it is often recommended that fine flour be put among shavings in a box smelling of honey as a pollen substitute in early spring. Bees certainly collect this flour, but scientific opinion is that it is useless, because it consists almost solely of starch which bees cannot digest. There is usually an abundance of natural pollen stored in the combs and any super combs containing much pollen should be given to the bees in autumn.

Feeders must always be made inaccessible from outside the hive, otherwise wasps and robber bees will become a perfect nuisance. If there is any doubt about the ability of outsiders to get through the roof, another cover-board or a sack should be put over the feeder.

It is also essential, especially in autumn, to feed only after sunset, or robbing may soon be started.

ROBBING

After the honey-flow ceases and until the weather becomes too cool for bees to go abroad, they prowl about in search of sweets. At this time they try to raid honey stores and sweet shops in search of anything of a sugary nature, paying special attention to other hives. Not only bees, but wasps persistently try to enter hives and only strong colonies are able to keep them out. They all become very much on guard and bees which are gentle as flies in the summer are often quite ferocious in autumn.

The bee-keeper must be prepared for this habit and not only take every care to reduce its effects to a minimum, but must himself avoid doing anything to start robbing and fighting. Honey or syrup spilt around the apiary is a fruitful cause of it and great care must be taken during feeding operations not to splash syrup about. Feeding should never be attempted until bees have ceased flying for the day and, if any syrup is spilt, it should be wiped up carefully or covered with dry earth. Above all, hive roofs must be carefully replaced after removal and no cracks should be left for bees and wasps to creep through; it is amazing how soon they discover the smallest passable crevice.

I think robbing, whether by bees or wasps, is most often caused by one of two things: removing honey late in the season, or careless feeding.

While the honey-flow is on, hives swimming with nectar can be opened and left exposed without fear of interference, but when there is little to be had in the flowers, an open hive will be invaded before one can say 'Jack Robinson'. Only experience will teach the beginner when the flow is over, for it is impossible to forecast it or fix a date by the calendar. In some years it is fierce and short, in others it continues gently for week after week.

Particular care must be taken of small nuclei and colonies below full strength. Strong ones mass a guard at the entrance, but weak ones must have the doorway reduced, so that only one bee can pass in or out at a time. If a hive is attacked, prompt steps must be taken to assist it. A piece of glass propped up before the reduced entrance is helpful, for robbers keep trying to get through it, while the rightful occupants go round the side. Perhaps the best of all plans is to stuff grass into the entrance and leave it till the evening. The home bees will have to cool their heels outside until flying has stopped, when they may be allowed to enter. Next day, a watch must be kept in case the raiding is renewed.

If matters go so far that robbing and fighting break out and become general, the garden hose should be brought out and used freely till the combatants are sobered down. Cloths sprinkled with carbolic hung in front of a hive will keep robbers away and wet sacks are also a good deterrent.

If it can be ascertained that one particular hive is attacking another, it is a good plan to change places, putting the robbed hive on the stand of the robbers' and *vice versa*. Thus the robbers are baffled and the hives made more equal in strength.

The great rule is to keep all colonies as strong as possible and if this is done there will rarely be this kind of trouble. Nuclei must always be carefully guarded and it is as well, if possible, to get them up to strength before the honey-flow is over. The most vulnerable stocks are those which are queenless or suffering from disease. Providing a queen is the remedy in the former case. If that cannot be done, the stock should be united to one with a queen, for it is impossible to maintain its morale otherwise. Colonies suffering from disease must be dealt with according to circumstances (see the chapter on diseases). The serious side to this matter is that a hive robbing an infected stock will take the germs home.

PREPARING FOR WINTER

Having satisfied himself that the hive contains a good population of young bees and an ample supply of food to last the winter and early spring, the bee-keeper's next task is to put the hive in the best possible condition to withstand whatever weather the winter may bring forth. Ideas on this subject have fluctuated a good deal. At one time it was thought that the more thoroughly insulated from cold, the better bees would winter, but this system of packing has now been thoroughly discredited. It has been found that even in the most severe winters, bees have survived in places where the temperature was little above that of the outside air, while those in hives almost hermetically sealed perished.

Although the individual bee is not able to survive a freezing temperature, the cluster—consisting of some thousands—can maintain temperature automatically by expanding and contracting. When the weather is mild, say between 40°–50°F., the cluster is only lightly held together, but as the cold increases, it contracts until it occupies only a small part of the former space. In milder conditions food is taken and this helps to carry the cluster through the colder periods.

VENTILATION

Ventilation cone

The real danger in winter is not cold but damp and bad ventilation. Bees, like animals, require oxygen and this can only be obtained when air can circulate freely in the hive. If the top covering is quite impervious and the entrance at the bottom very small, air will circulate very slowly and will soon become charged not only with carbon dioxide but with moisture exhaled from the bees. In such conditions the flame of life burns feebly and soon flickers out. In spring the bees are found dead and the combs covered with mould.

There are two good ways of ensuring ample ventilation in the hive. One is by having a wide and deep entrance at the bottom and a covering which is not quite impervious to air and moisture. If quilts are used, they should be of porous material, which will allow accumulated moisture to pass off steadily and constantly upwards and above them there should be sufficient space to receive this moisture, with ventilation holes through which it can pass right out of the hive. If wooden covers are used, the piece of glass usually laid over the feed-hole in summer should be replaced by perforated zinc and, by means of strips of wood placed inside the corners, the roof should be raised sufficiently to permit air to circulate freely over the board.

Naturally, it is important that moisture should not enter the hive from without. Nothing could be much worse than rain leaking through the roof, yet often insufficient care is taken to prevent it. Flat roofs do not give much trouble, as they are generally covered with one piece of zinc or bitumenized felt, but gable roofs often develop leaks through which rain can trickle into the hive.

For the purpose of ventilation, the bottom entrance can scarcely be too big, but there are dangers to be avoided. Mice are prone to take up their abode in beehives if the entrance will admit them. They consume the comb to make room for a nest and even if they do not cause the death of the colony, they spoil valuable comb and make an unholy mess. They must be kept out and as they can creep through an incredibly small space, the ordinary entrance is too deep. The most satisfactory plan is to open the entrance to its full extent and cover it with a piece of excluder zinc; through this nothing larger than a worker bee can pass, while plenty of space is left for air to enter.

The other method of wintering, in great favour in some places where the winter is very severe, originated, I think, in British Columbia. It is to close the bottom entrance and make one at the top. This system allows the warm moist air to pass off immediately and fresh air to flow in. This upper entrance is also less accessible to mice.

If the apiary is in a garden, reasonably sheltered from high winds, no further precautions are needed for safe wintering, but in exposed places it is necessary to guard against the possibility of roofs blowing off or even the whole hive being blown over. Roofs can be secured by a hook and eye, or they may, in British National, Langstroth and M.D. hives, be twice the normal depth so that they cannot blow off.

If there is danger of the hive blowing over, a wire rope can be passed over the roof and attached to pegs in the ground. If fibre rope is used only one side should be pegged. The other should be fastened to a brick to allow for contraction, which will merely lift the brick.

These preparations for winter should be completed by the end of October, so that the bees can be left undisturbed till March.

If snow falls or seems imminent, the entrance to the hive should be shaded, because the intense light from snow-covered ground is reflected into the hive and tempts bees to fly. They seldom get back, but alight on the snow and perish. A board propped in front will avert this and also keep snow from blocking the entrance.

Langstroth hive prepared for winter with a mouse guard covering the entrance and the roof securely weighted

SELLING AND STORING HONEY

Not the least attractive aspect of bee-keeping is that it is a summer occupation and for the five months from November to March there is nothing to do outside but take an occasional look round to see that everything is normal. What little there may be to do can be done indoors—repairs to hives, preparing frames and supers for the next season and so on. The enthusiast will also find a large field of literature open to him. Hundreds of books, ancient and modern, deal with the complexities of bee life and practical bee-keeping. You can never know too much about bees, and the oldest and most experienced bee-man always gives a welcome to a new book on his subject.

One matter may take up a fair amount of time during winter, for if the crop has been a good one, there is the work of putting it into attractive form for disposal. Even the owner of only one or two hives may quite well have a hundredweight or so of honey which he cannot consume at home, and he should know how to market it satisfactorily.

Whenever it is possible, honey should be put in the final containers immediately after extraction. The less it is handled the better, for every time it is poured from one vessel to another some of the delicate essence is given off, especially because, after a time, it sets solid and cannot be transferred without being heated—a process which accelerates the loss of aroma and flavour.

DISPOSAL IN BULK

Those who do not want to undertake retail marketing can dispose of their surplus in bulk to various firms who repack it for retail sale. All that is necessary is to run the honey from the storage tank into tins sold for the purpose. These hold 7, 14, 28 or 56 lb, but by far the most useful package is the 28 lb. Some of these firms will send tins if arrangements are made in advance, but they are always essential for handling and storing large quantities, so a few of 14 and 28 lb size should be part of your equipment.

After standing a few days in the storage tank, during which air bubbles, scraps of wax and so on will rise to the top, the honey is drawn off into the tins and can then be despatched by road or rail. In the latter case it is advisable to pack the tins in a crate so made that the handle projects for carrying. This will usually ensure that the tin is not turned upside down in transit, but, of course, the lid must also be firmly secured.

IN JARS

Many people will be able to find an appreciative market in the neighbourhood for honey in jars, so a certain amount, if not all the crop, will need packing in this way. In Britain honey was invariably put into a tall narrow jar containing one pound. It had the advantage of looking most alluring, as the narrow jar made it more transparent and bright in appearance, but it was an awkward vessel to get the honey from, especially when it had crystallized. This jar has gradually been superseded by the 'squat' jar, which is little more than half the depth, but much wider. Various patterns have been introduced from time to time, but the one which has become the standard pack is the British Ministry of Agriculture design; it is easy to fill and just as easy to empty. It has a screw cap which fastens readily, but is virtually airproof.

Whatever vessel is used for honey, it is most important that it should be airtight. If moisture-laden air can reach it, honey will absorb water and ferment, so it must have an impervious cover. Nothing is better than the screw cap, though what is known as the 'bayonet catch' serves very well. If in an emergency—as when an unexpected heavy crop fills the standard jars—common glass jars have to be used, they must be covered with something airproof. Plain paper alone is quite useless, but if used with a waxed paper inner cover, any paper will do. It is also possible to buy metal covers to fit glass jars, and these are quite satisfactory; tying down paper covers takes a lot of time and never looks so neat as a metal cap. Waxed pulp pots are also sold for packing honey and these have the advantages of lightness for transport and being unbreakable, but they have not the same appeal as glass, which shows the full beauty of honey. However, they come in handy for dark-coloured honey, which does not have the same æsthetic appeal as the light or golden varieties.

IN CANS

Honey packed in cans for wholesalers need not be very finely strained. It will all be melted, blended and strained by the packers, but before packing in jars all impurities must be strained out. In the usual storage tank there is a loose top, holding at least 28 lb of honey and this has a coarse grid at the bottom. Below this is a flange with rolled edge and, to strain honey properly, two thicknesses of butter muslin should be tied securely over this flange. Since it will need to be renewed from time to time, as it becomes clogged, it should be tied in a bow, so that it can easily be undone.

FILLING THE JARS

For convenience in filling, the tank should be raised on a bench, at which one can sit comfortably, with room for both empty and full jars. Needless to say, the jars should be thoroughly cleaned before filling. They are often very dirty when supplied and the best plan is to soak them in a tub full of water, swirl a bottle brush round inside each and stand it to dry in the sun, or before a fire, as there must be no moisture inside when the honey is run in.

Jars are filled very quickly at the valve and must be constantly attended to avoid running over. They should be filled a little above the ring which marks the correct height and the caps put on lightly at first. They should stand in a fairly warm place for twenty-four hours to permit any scum to rise. This can then be removed with a spoon with any excess honey above the ring. By the time all jars are done there will probably be one or two more filled with this excess and after standing a while, they can be treated like the others. This extra care is well worth while, as it ensures even filling and the honey sets without a crust on top.

Honey spoon

As the jars are completed, the caps should be screwed down tightly and a label put on. There is a large variety of labels on the market, and the beekeepers' associations also usually supply one to their members. Honey which is to be sold in shops must have the net weight and the raiser's name and address on the jar. If there is not enough to warrant special printing, a rubber stamp can be had fairly cheaply.

Gummed labels can be stuck on readily if the jar is wiped with a damp cloth and the label pressed on the side. A final wipe with a dry cloth completes the work.

STORING

Although the jars are as near airtight as possible, it is wise to store them in a dry place where the temperature does not fluctuate too much. Nothing is better than a cupboard in the kitchen.

Sooner or later, the honey will set hard and crystallize in the jars, the time depending on the kind of honey. Fruit blossom and clover honey are rather slow to crystallize, while that from charlock and other crucifers sets so quickly that it is often difficult to deal with without warming it.

My own view is that honey allowed to crystallize naturally without interference is by far the best, but some people like it to be soft and creamy. Naturally granulated honey often shows what are called 'frost marks' on the side of the jar. They are caused merely by contraction during crystallization, and have not the least effect on the quality of the honey. Indeed, they are a guarantee that the honey has been bottled without heating, but some large bee-keepers contend that the public eye should be considered and anything which makes people doubtful avoided.

Those who agree with this view must go to more trouble. After straining, the honey should be drawn off into cans and allowed to crystallize. It is then partly re-liquefied by standing the can in hot water, and then bottled. It need not be entirely liquefied, but softened sufficiently to permit it to be stirred until it is smooth and creamy and can be poured into jars. After this it will not set hard again, but will be easy to take out with a spoon.

If you have too much honey to bottle at once, it must not be left in the storage tank to granulate, or it will have to be dug out laboriously. If run into cans it can, as I said, be melted in hot water, or by standing on the hot tank in the bathroom. It is best to melt honey slowly at a moderate temperature and in no case should it be allowed to go beyond 140 °F. I strongly advise bottling at once, if it is at all possible, to retain the best flavour. Care in bottling and neat labelling will always pay, especially if one is asking shopkeepers to show your honey. In these days of factory-packed goods, attractive packing is the only way to secure attention from buyers.

I advised that sections should be stored in cans after being roughly cleaned of wax and propolis. In this way they will usually remain liquid for months, but since few people care for crystallized sections, they should be disposed of as soon as possible—they are particularly appreciated at Christmas. At one's leisure they should be cleaned more thoroughly and protected from dust and flies, not to mention bees and wasps. Printed cartons can be bought for this purpose.

Care expended on packing and labelling jars or sections will always be worthwhile, as it will secure favourable notice at sight and a good reputation for the product, not only among consumers, but what is more important when there is a good deal to dispose of, among the retail shopkeepers. The standard of packing of all kinds of food has reached such a high level that no shopkeeper will display anything slovenly or dirty in appearance, however good the contained article may be.

Careful packing is necessary when jars of honey or sections are to be sent by post or rail. Each should be separately rolled in corrugated paper and the whole put into a box large enough to allow plenty of straw, hay, or other suitable material to be packed all round them.

To ensure a good market, one must take care that the full flavour of the honey is retained. It should not be extracted until sealed by the bees and extraction should be done in the driest possible atmosphere, as quickly as possible. If it is not practicable, as is best, to bottle it immediately, it should be stored in 14 or 28 lb cans, so that it can be re-liquefied without overheating. Whether in cans or bottles, it should be stored in a dry place with a fairly even temperature somewhere between 40° and 60°F. Treated in this way, it may be trusted to remain in perfect condition almost indefinitely, so that it can be marketed all the year round.

7
Honey
Sources

Even in his first season, the observant bee-keeper will not fail to notice great variation in the bees' activity. One day they will pour from the hive incessantly. The next perhaps only an odd bee or so is seen outside. Even on any one day there may be little activity, except for a single hour, when it seems as though foragers could not get out fast enough. What is even more puzzling, is that sometimes one stock will be working hard, while others are quite sluggish.

These fluctuations are mainly due to weather conditions. Bees are proverbially creatures of the sun and are always more active on bright days than on those which are cold, wet or windy. This, however, is by no means a complete explanation, for there are often days of brilliant sunshine and comparatively little coming and going from the hives. On the other hand, one often sees signs of a strong honey-flow on summer days of continuous rainy drizzle.

WEATHER, FLOWERS AND NECTAR

The response to warmth and moisture is not alike in all flowers. Just as some seeds germinate at temperatures as low as 40°–50°F., while others require 80°–90°, so some flowers yield nectar freely when the mean temperature is no more than 45°, while others require a day temperature of 70° before they begin to secrete. Generally speaking, these nectar-yielding minima correspond with the time of year at which the various plants blossom. The fruit trees, blooming in spring, afford nectar freely at much lower temperatures than summer-blooming plants like clover.

This difference in nectar-yielding capacity is also affected by the nature of the soil, more especially in the case of herbs. This is to some extent due to the mineral constituents of various types of soil, but is chiefly because light soils warm up more rapidly than heavy land. Subject to the modification caused by these variable conditions, there is a regular succession of flowers which are known to give nectar freely and are the foundation of the honey crops produced in Britain.

This succession begins with the willows, the first being the sallow or goat willow which blooms about Easter and is often used for church decoration under the name 'palm'. On sunny days, sallow bushes are thronged with bees if there are any hives anywhere in the vicinity. The white and crack willows follow and are less noticed, because of their height, but bees are active among them and often bring in large quantities of honey for the time of year.

FRUIT, FLOWERS AND BEES

Fruit blossom begins at much the same time, the earliest being almond and peach. These are not very common, but are greatly favoured by bees, who will enter peach houses in large numbers if the ventilators are left open. Plums do not seem quite so attractive, but cherry, pear and apple trees are great favourites and in good weather a strong stock of bees will gather considerable quantities from them. Apples give amber honey of good quality and since cross-pollination is essential for a good crop of fruit the benefit of keeping bees in orchards is twofold.

Currants and gooseberries are also well worked, to their own and the bees' advantage. I have known a gooseberry plantation which had previously yielded very little fruit, bear most bountifully when a hive was set down near it. Hawthorn is a common and long-blooming tree, rather uncertain in nectar yield, but in some years giving a good crop, while the insignificant blossom of holly also yields freely and constantly. Maples, *Acer* spp., are the last of the early-blooming trees to provide nectar in quantity. Sometimes such a roaring of bees can be heard in the sycamore maples that it sounds as if they are swarming.

In April or earlier, dandelions gild the meadows and one can see by the big loads of pollen brought in, how valuable this handsome weed is to bees. It is probably the major source of pollen at the time brood rearing is at its height, but it also yields abundant nectar and if its blossoming time were later, when there is a stronger force of foragers, it would be an important source of surplus honey.

During April and May, cabbages bloom in gardens and allotments and give nectar freely, though unfortunately this honey crystallizes very quickly and coarsely so that it often cannot be extracted from the comb.

The next plant of importance is the raspberry, which has a long blooming period, but is at its best about the beginning of June. Where this fruit is grown in quantity, it is most useful for tiding over the June gap during which bees are often hard pressed to get enough for daily needs and must fall back on reserves. Field beans yield nectar plentifully and consistently and will often add substantially to the surplus. I have known a stock gain 30 lb in the week when such a field was at its best.

The first of the fodder crops in Britain—sainfoin—opens the main honey-flow about the end of May. This yields a bright yellow honey of delicious flavour. Sainfoin is usually cut for hay at the peak of blossom, but blooms again in August, and while it is in bloom, bees generally neglect other plants. Like its relatives the clovers, it flourishes best on chalk soils and is not so plentiful on clay lands.

From the beginning of June the clovers come successively into bloom, the first being crimson clover, usually known as Trifolium. This is annual and needs to be sown every year. Yellow trefoil or medick and yellow suckling clover come next, but these are not so much favoured by the honeybee and are more frequently used by the small solitary bees. Alsike clover is a useful variety on low-lying damp soils. Red clover is the species most useful to farmers and it contains also the largest amount of nectar, but its tubes are normally too deep for the honeybee so that it is really the perquisite of bumble bees. Hive bees rarely get anything from the primary crop, but after this has been cut for hay, the plant blooms again in August and the flowers are smaller and more accessible. In hot weather enormous yields sometimes come from this source in a few days.

Outshining all other clovers is the white, creeping, or Dutch clover which grows so freely, not only where sown in permanent or temporary pastures, but

Italian bee collecting pollen from red clover blossom. Adequate pollination of red clover often doubles the yield of the clover seed

by roadside and waste places everywhere in its wild form. It is particularly abundant on chalk hills like the South Downs of England, and it is in such districts that the largest and purest crops of white clover honey are gathered. This honey is highly esteemed for its delicate flavour and attractive amber colour.

Alfalfa, *Medicago sativa*, is another leguminous plant which is a good nectar yielder on suitable limestone soils. In the United States it ranks among the best honey plants, but it does not figure very largely in the British bee-keeping scene, mainly because it is rarely allowed to flower. It is grown for green feed and cut three or more times a year. If allowed to flower it becomes shrubby and unsuitable for making hay.

Vetches or tares are sometimes grown in large quantities for seed and these have the property of producing not only floral, but extra-floral nectar, from the axils of the leaves, so that nectar sometimes comes from them before they bloom.

During the first half of July the main honey-flow reaches its peak with the blossoming of the basswood, whose scent often pervades the air during hot summer days. In favourable weather, large quantities of honey are gathered from these trees and they are particularly valuable in and around towns where there is not much clover. The humming of bees in basswood is one of the most joyful of summer sounds and it is specially gratifying to the bee-keeper, for it foreshadows a harvest of richly flavoured honey.

HONEYDEW

After these two important crops, there are in most districts few plants which can be relied on to add much to the surplus, but at this time, especially in very hot weather, bees sometimes collect large quantities of 'honeydew'. This is not true nectar, but a sugary exudation from the leaves of many plants, notably basswood and oaks, said to be induced by the attacks of aphis and other piercing insects. Honeydew is a very doubtful blessing to the bee-keeper. It varies a good deal in colour and flavour, but is always darker than floral honey and sometimes as black as soot. Moreover, it often contains sooty particles, which may either be actual soot deposited on the leaves and mixed with the sugar, or the product of a fungus which is liable to attack the sugar. The bees appear to like this substance and, indeed, its flavour is often quite agreeable, but it is not attractive in appearance and is considered to be liable to cause dysentery in bees which have stored much of it to winter on.

Not much can be done about it, but when extracting, the presence of this in the combs can usually be detected by holding them up to the light; such combs should either be kept in store for feeding back to bees in spring—not in autumn—or extracted and kept separately for use in cooking.

In August there may, as I have hinted, be a fair amount of surplus from red clover or sainfoin. Brambles also yield fairly consistently and in some places the woodland willow herb or fireweed adds its pale honey to the quota.

Only one other substantial source of honey is likely to be met with in Britain and that is confined to peaty and sandy soils, such as the extensive moors scattered about the British Isles, but more especially associated with the broad acres of Yorkshire and the Scottish highlands. In such places, the characteristic plant is heather which, under favourable conditions, gives immense quantities of honey. There are several species, but the ling, bell heather and crossleaved heath are universally met with on such moors. The last two produce dark honey, not otherwise different from what the Scots call 'flower-honey', but ling produces a peculiar jelly-like honey which rarely if ever crystallizes and has to be extracted from the comb by special technique. Heather blooms from mid-July to late September. In Scotland it is looked upon as the main crop and special steps are taken to secure the largest possible harvest from it.

Although the above are the main sources from which honey is obtained in large quantities, they by no means exhaust the list of nectar- and pollen-producing flowers. Hundreds of plants have been listed as having been visited by the honeybee, so that mere enumeration would take up a lot of space.

Among trees, one may mention the alder, hazel and elm, which do not yield nectar but, flowering very early, produce a welcome supply of fresh pollen. Later comes horse chestnut, whose rich crimson pollen is unmistakable. The yew and box are also valuable sources of pollen. In June a well-grown tree of black locust, *Robinia pseudoacacia*, will produce large quantities of its white flowers and swarm with bees for a few days.

Even the vegetable garden and arable field produce honey. The asparagus plant, though its flowers are inconspicuous, yields much nectar which bees rarely fail to find, while everyone must have noticed how fond they are of onion flowers and wondered perhaps whether the resulting honey has an onion flavour. Here it is rarely gathered in sufficient quantity to affect the general stores, but in some places where wild onions are common we learn that, although strong smelling when first extracted, the onion flavour disappears after a few months' storage.

All the labiate herbs—balm, lavender, marjoram, mint, rosemary, sage and thyme—are good sources of richly flavoured honey and I have already mentioned the cabbages which are sometimes allowed to go to seed and afford a long income for the bees.

On arable land that troublesome weed, charlock, makes some amends by providing plenty of nectar for the bees, while its near relative, mustard, yields bountifully when grown in sufficient quantity. Buckwheat again, is a high-yielding honey plant, but this is not now a very widely grown crop.

Purely wild plants yielding nectar and pollen in variable quantities are innumerable and the following list contains only a selection of common ones—bird's-foot trefoil, bluebell, buttercup, corn sowthistle, crane's-bill, dead nettle, field bindweed, figwort, loosestrife, mallow, mustard, parsnip, ragged robin, scabious, thyme, thistles, viper's bugloss.

PLANTING FOR BEES

New bee-keepers often write to the journals asking what they should plant for their bees and although it is not profitable to grow plants *only* for this purpose, there is no reason why garden flowers, grown solely for the pleasure they give by their beauty and fragrance, should not include some of those most favoured by bees. No music is more delightful to the bee-lover than the hum of foragers in a cotoneaster bush, or a bed of mignonette. I give, therefore, a list of such garden subjects as I know from experience to be extremely attractive to bees.

Shrubs Berberis, buckthorn, buddleia, cotoneaster, erica, genista, ribes, shad-bush, snowberry, veronica.

Perennials and Biennials Anchusa, arabis, aubrieta, campanulas, Canterbury bells, crane's-bill, centaurea, forget-me-not, French honeysuckle, globe thistle, hollyhock, linaria, mallow, asters, nepeta, salvias, sidalcea, sedums, veronica, verbascum, violet, wallflower.

Summer Bedding Plants Dahlias, fuchsia, heliotrope.

Bulbs All the early-flowering bulbs are visited by bees, especially crocus, hyacinth, narcissus, snowdrop.

Annuals There is a wide choice of these, but the following are outstandingly attractive to bees: borage, cornflower, clarkia, gilia, limnanthes, mignonette, phacelia, poppy, scabious.

Generally speaking, it is not much use trying to improve a poor bee-keeping locality by planting specially for bees. The causes which make a poor district are usually radical—something in the soil which inhibits nectar secretion. On the other hand, when soil and climate are favourable, it is unnecessary to plant anything, for natural growths will generally provide all that the bees are likely to be able to use.

TREES AND SHRUBS

Several other ways in which dual purposes can be served will occur to the experienced gardener. When trees or shrubs must be planted to serve as windbreaks, there is no reason why nectar- or pollen-producing ones should not be selected. It is worth remembering perhaps, that there is a great demand for holly at Christmas and that the most berries are likely to be produced on trees freely visited by bees. As holly makes an excellent windbreak and an impenetrable hedge, it can thus be made to serve four distinct purposes.

Speaking of holly reminds one that many flowers are grown for market. Most of them are not much use to bees; but lavender is greatly favoured, as are cornflowers and scabious. Single dahlias are very popular, both with bees and housewives, for decoration.

Caucasion bee collecting nectar from holly (*Ilex* sp.) in May in Missouri. This is an important source of nectar during the build-up period, and the shrubs are found in a large area of the Southern States

Concern is sometimes felt about the supply of pollen in early spring. On the whole this is more imaginary than real, but in very bare open country, excellent for bees in summer, there is often a dearth of early supplies. In such places, dwarf willows, like *Salix aurita*, may be planted on waste land, but for the earliest supplies nothing surpasses hazel. Cobs and filberts may also be planted, not only in hope of a profitable return from the nuts, but as windbreaks. These will supply the bees with lavish quantities of pollen.

It is rarely safe to rely on later forage for filling up winter stores, but mustard sown in August for sheep feed or green manuring will bloom freely in October and provide bees with plenty of work if the weather is genial. Which reminds me that I have not mentioned the very last source of nectar in the year—ivy. When this is in bloom during October bees crowd to it in tremendous numbers on fine mild days, almost as though the season were beginning, instead of closing, for them with one great glorious feast.

MAJOR NORTH AMERICAN HONEY PLANTS

The plant genera and species from which honeybees obtain nectar in North America run into many hundreds. Some are of local significance, while the majority contribute little or nothing to the final yield of honey, although they are important to the well-being of the colony. Various authorities estimate the number of species producing an exploitable crop to be under one hundred.

Dr Eva Crane lists one hundred and fifty genera and species of world-wide significance and includes over sixty that are found in North America in *Honey: a comprehensive survey*.

F. A. Robinson and E. Oertel (*The Hive and the Honey Bee*, 1975) give about two hundred important nectar and pollen plants and indicate the regions of North America in which they are found. The table below lists the seventeen plants they regard as of major significance.

Plant	Nectar season	Regions and Notes
Alfalfa (*Medicago sativa*)	late July to early Sept.	Southwestern, Western and Northeastern states. Grows best on well-drained soil
Aster (*Aster* spp.)	autumn and winter	Florida up to Canada and Western states
Basswood (*Tilia* spp.)	late June and July	most areas; produces light-coloured honey of distinctive flavour
Black mangrove (*Avicennia nitida* Jacq.)	June and July	restricted to coast of Florida and Southern states
Buckwheat (*Fagopyrum esculentum*)	late July and August	now limited; once popular in North-Central and Southeastern states and Eastern Canada
Citrus (*Citrus* spp.)	spring	Florida and South. Includes tangerines, oranges, grapefruits etc.
Clovers (*Trifolium* and *Melilotus* spp.)	mid-June to late August	most important honey plants, mainly in North-Central states
Cotton (*Gossypium* spp.)	summer	South and Plains region
Fireweed (*Epilobium angustifolium*)	mid-summer to autumn	Northern states, Canada and Pacific Coast
Gallberry (*Ilex glabra*)	May to mid-June	common to the Southern states; a major source in Georgia and Florida
Goldenrod (*Solidago* spp.)	June to Sept.	widely distributed in USA; major source of supplemental nectar
Sage (*Salvia* spp.)	June to Sept.	three species of major importance in California
Saw palmetto (*Serenoa repens*)	April and May	Southeastern states only
Sourwood (*Oxydendrum arboreum*)	June and July	North-Central and Southeastern states, especially Eastern states
Soybeans (*Glycine max*)	summer	Central-Plains states, Arkansas and Missouri
Spanish needles (*Bidens* spp.)	late summer or autumn	common in Eastern states and parts of Canada
Tupelo (*Nyssa* spp.)	April and May	Southeastern states, especially Florida

8
Spring Management

Whether or not the beginner has secured a crop of honey in his first season, he should, by the time the second opens, have become used to handling his charges and have a fair idea of their habits. After the long confinement of winter, he will look forward eagerly to seeing the bees on the wing again and perhaps be tempted by a fine day to open the hive and see how things are progressing.

Such days not infrequently occur as early as February, but that is much too soon to interfere. No good can be done by it and there is possibility of serious harm. For some reason which has never been fathomed, a stock that is opened at this time may possibly 'ball' its queen. It does sometimes happen at other times, but at this stage, when there is only a little brood, a stock which has been disturbed by smoking seems to become panic-stricken and a lot of bees rush for the queen and surround her so closely that, if she is not quickly released, she is suffocated. It has been suggested that this is because of the colony's extreme anxiety to protect her at this critical moment. At any rate, it seems to happen most frequently at a time when there are no young bees but only old over-wintered ones. If it is really necessary to risk an examination at this time, a sharp lookout must be kept and if the bees run about hurriedly, the ball, about the size of a walnut, will be seen rolling on the floor. Prompt action must be taken, the best plan being to pick the ball up and drop it into water, when the bees will break away and the queen can be picked out. If there is no water handy—you have been warned that there should be—smoke the ball heavily until the queen is released. This is not enough, for as likely as not, she will be balled again when you have gone away. She must be caged on a comb for a few hours and released later on. It is always desirable to have on hand at least one queen cage, when stocks are manipulated. The simplest is the domed 'pipe cover' cage, which can have its open end pushed into a comb, picking out a piece with honey, so that the queen has access to food if her release is delayed.

AN EMERGENCY CAGE

In emergency, a matchbox is a suitable temporary cage. If opened to its full extent, it can be slipped over the queen and carefully closed. It is quite a good idea to have a special matchbox with a piece of glass in the cover. When the queen is covered by the box, she makes for the light shining through the glass and this allows you to close the box safely. The box can then be laid over the top of a frame and opened not more than $\frac{1}{8}$ in. The bees can thus get in touch with the queen without being able to injure her and they will bite away the edge of the box until she can get out, by which time normal relations should have been restored. A balled queen rarely escapes without some damage, like frayed wings and her hair torn off and one must not be surprised if she is superseded.

In any event it is rarely necessary to open hives so early in the season. If the roof is removed and the hand laid on the centre of the cover, the amount of warmth will indicate whether breeding has begun and by lifting the hive from its stand, it is easy to tell whether there is enough food. If there is any doubt about this, feed by putting on a tin of syrup, wrapping it up so that too much heat does not escape. It is when breeding has well begun that most warmth is required.

WATERING

At all active times, except during a very heavy honey-flow, bees collect much water, especially in spring when it is needed to dilute the winter stores of thick honey. They can be seen congregating in large numbers at the edge of streams and pools, sucking up water from the ground or leaves. In towns they often go to house drains where, perhaps, a dripping tap keeps a constant supply of moisture present. This is dangerous, for a sudden rush of water will wash many bees down the drain. It is just as well to cover such places so that bees cannot reach them, and to provide a special supply of water for them. There are many ways of doing this. In a small apiary, one or two glass jars can be inverted over a board with grooves scored in it, so that just enough water seeps out to keep the board moist. If placed in a sunny spot, the bees will soon take to it, especially if a little sugar is put in the first supply. Some use salt, which seems equally attractive. In a larger apiary, a tub may be supported on a stand and a tiny hole bored at the bottom to allow a continual drip onto a board placed below. A bird bath filled with pebbles for the bees to stand on will also serve.

Unless it is suspected that something is wrong, April is quite soon enough to open hives; if you wait till the apple blossom is well out, it will be about time to make the first inspection. If the colony is full of bees, nicely compacted on the combs, it is not necessary even to take a comb out. By blowing a little smoke between the centre combs, it is easy to see whether there is brood and that is all that matters. If the colony covers eight combs or more, it should have another standard brood-chamber. If only six or seven are occupied, a shallow super will serve. Opinions differ as to whether this should be put above or below the other, and it does not make a great deal of difference, but in the natural way bees work upwards to where their stores are placed in winter and downwards in spring as brood increases, so that to put the new chamber underneath seems better. The drawback is that in order to see how it is progressing, the other has to be lifted. If possible, this addition should be of built-out combs, as foundation is not so nicely worked on early in the year. It is always better to get combs built in the height of the season.

A practice at one time recommended was 'spreading brood' by putting a frame of foundation in the centre. It is doubtful if this increases the amount of brood produced and there is danger that if the weather turns cool some will be chilled, owing to the inability of the bees to cover a larger area of the comb in spring.

93

CLEANING THE HIVE

Before adding the new chamber, it is well to scrape off any brace comb built on the old combs—the work of a few minutes with the hive tool—and to clean the floor board. It is best to use a fresh spare board, lift the brood-chamber onto this and clean the old board thoroughly in readiness for another colony. If there is no spare board, the brood-chamber can stand on the upturned hive roof while the floor is cleaned. A reversible board can be turned over to provide a clean floor.

If a colony at this time contains no more than five bee-covered combs, it should be examined comb by comb to make sure the queen is present and laying, and to satisfy yourself that the brood is healthy. For signs of disease the reader must refer to the chapter on that subject, but here it may be said that sealed brood, which is not quite even throughout but has jagged holes here and there, or cappings which have sunk, should be regarded with suspicion. If there are, say, two combs of healthy brood, the colony can be covered up and left to expand for a few weeks. Sometimes a comparatively backward lot in early spring will build up to good strength by the time the main flow arrives. Judicious feeding may supply the needful stimulus.

If there is only one stock, it must be built up as well as possible, for at this time of year nothing else can be done, but when there are several stocks, it is sometimes worth while to unite two together. Considerable discretion must be exercised in this matter and only experience will tell whether a particular lot is likely to build up well, whether it might usefully be added to another, or if it might not be better to destroy it. On the whole, I think that, however small the colony may be, if it is compactly clustered, has brood proportionate to its size and is active even though with few foragers, it is worth helping on by some encouragement. Empty comb should be removed and a division board put close up to reduce the space and, of course, a good supply of food should be given.

If the colony is not merely weak in bees, but scattered about the combs inside the hive and sluggish outside, it is probable that there is something wrong, perhaps one of the less spectacular diseases like paralysis, and it is wiser to destroy such a lot forthwith.

UNITING STOCKS

It may be that a stock fairly strong in bees has no brood whatever. Search must then be made for the queen. Perhaps one is not found, or being found, is seen to be old, shrunken and frayed: it may be that there is brood, but only drone. In any of these cases, it is wise to unite the stock to another, having first removed the old worn-out queen.

There are two recognized methods of uniting stocks, but for both the preliminaries are the same. They must first be brought together by stages, if they are more than 6 ft from each other. Each is moved in the direction of the other a yard every day. When they have thus been brought alongside each other, the old queen, if any, is removed. If both colonies are weak, the best plan is to remove all empty combs, so that the occupied ones can all be put in one brood-chamber. Space the combs of the queenright one far enough apart to allow a comb to go between each pair. Now dust both lots of bees with flour, not too thickly, but enough to make them clean themselves. Insert the combs of the queenless lot in the spaces left in the other, close them up together and cover up. Some recommend that the queen should be caged, lest the strange bees should attack her, but I have never known this happen. It is important that a fine warm day be chosen for this operation.

In the other method, which is only suitable if both stocks are fairly strong, neither brood-chamber is disturbed, but one of them is covered with a sheet of newspaper, and the other put on top of this. In due time the bees eat their way through the paper and unite peaceably. After either method of uniting, the empty hive should be cleared away and the full one placed midway between where the pair last stood.

Bee-keepers of considerable experience are often concerned, on examining their bees in April, to find them weaker than when they first emerged from winter confinement, but if the facts are soberly considered, this is quite natural. From the time the colony enters its resting period in November, when the hive is, or should be, crammed with bees, the workers are not subjected to any strain or risk until the early flowers bloom. Older bees die off but there is not a very heavy loss of population in a healthy stock. When the first spring days bring snowdrops, crocuses and early blooming trees, the old bees begin to forage and the treacherous conditions of these early weeks take heavy toll of them. The queen will have begun laying, but the number of bees maturing is small and quite insufficient to replace losses. Hence it often happens that 'spring dwindling' is more serious when spring comes early than in years when hard weather continues into March. So long as there is a good queen, this dwindling need cause no alarm, for it will be made good rapidly when the batches of brood mature into adults.

STRONG FORAGING FORCE

The chief secret of bee management is to have the strongest possible force of foragers in the hive at the time the main nectar plants are in bloom. Left to themselves, bees build up their colonies on the flowers of spring, pass through the period of reproduction and then collect the stores which are to carry them through the rest of the year. There is really not a great deal the bee-keeper can do to accelerate this development. His part is to select the colonies which give best results, provide them with the wax foundation which enables them to build comb rapidly at small cost of food consumption and to see that *at all times, there is plenty of food* in the hive. Any period of shortage inevitably means slowing down of breeding and general development. Seasons vary so much that it is not easy to estimate in advance just when a particular crop will bloom, and the conditions which obtain in one district are often quite inapplicable to another. Hence the golden rule is to 'keep all colonies strong at all times'.

I suppose there are few parts of Britain where there is not some early nectar obtainable, generally from fruit and forest trees and these serve to build up the stocks for the main flow. In some places, where fruit growing is a dominant feature, there is a comparatively poor supply of nectar later, so that the fruit blossom must be regarded as the main source and steps taken to get stocks as strong as possible for it. By putting the hives right inside the orchards, much of the risk of exposure to strong winds will be avoided.

On the other hand, moorlands provide little or nothing before the heather blooms and the problem there is quite different. It is usually dealt with by taking the bees to the moors after the clover is over.

The bee-keeper must, therefore, study his own district and form an idea of the time when the main crop may be expected; then he must, by every possible means, contrive to have the foragers ready for it, bearing always in mind that it takes about six weeks from the time the egg is laid till the worker is ready for the field.

9
Record
Keeping

If the apiary contains only one stock, 'the bees' is sufficient description for it, but when the number increases, it is necessary to adopt some means of distinguishing one from another, for it is impossible to deal with anything intelligently if it cannot be identified in some way. So, as dairymen give delightful names to their cows, and poultrymen ring their birds' legs, the bee-keeper numbers his stocks, according to some system of his own, which may be crude, elaborate, haphazard or purposeful. It is a subject which has a good deal more in it than first appears, and is worth discussing in detail.

First, it should be clear that the number refers to the colony in a hive and not the hive itself. I can conceive of no object to be served by numbering a hive, except to identify the bees in it, so that *ipso facto*, the number refers to the bees and not the building. This suggests at once that it is bad practice to fix the number permanently on the hive by painting or writing, or even by screwing numbers on, because not only do we sometimes find it desirable to transfer bees from hive to hive, but the identity of a colony often changes, as after the issue of a swarm. This again causes us to consider what constitutes the identity of a colony, and we need not hesitate to say that it is the queen at its head. The moment the queen is changed, the colony's constitution is more or less completely altered, so that in essence the number refers to the queen.

Numbers should, therefore, be movable. Appliance makers and iron-mongers supply them in various styles, but a piece of zinc about 2 in square with the number painted in white will do. One tack driven through a hole in the top will serve to affix it and it can be instantly removed. It can be put where one fancies, but usually the back of the brood-chamber is best.

A SUGGESTED RECORDING METHOD

I change my numbers every year according to a system of my own, numbering colonies according to their performance the previous season. If I get new stocks or queens, I run the numbers on for the season, but if I raise queens, I number the nuclei the same as the mother, thus 1a, 1b and so on. At the end of the season when results are tabulated, the daughters are placed next in succession to the parent. If No. 1 has three nuclei, a, b, c, but proves to be only third in merit that year, mother and daughters will be 3, 4, 5, 6, or if the mother is done away with, the daughters are 3, 4, 5. There may be objections to this system, it may even be misleading as far as colony merit is concerned, but it is easy to follow and ensures a continuous record.

Every bee-keeper worth his salt keeps records, if only in his head, and if records are desirable, they should be well-kept. The usual plan is to have a card for each stock. I have seen many such, ranging from old post-cards to

specially printed ones, most of which consist merely of Colony No......
Date......Notes...... all of which being self-evident, the printing seems a
little superfluous. My own records are kept on a plain sheet of 8 vo paper and
I cannot do better than reproduce an actual specimen.

1951		QUEEN 1949	1950			8
			2	Surplus taken		66
May	1	8/9 Supered 10 B/S		left		20
,,	30	Super half full and sealed				
June	2	10 shallows added				86
July	20	Supers removed. Extracted 66 lb		*Less*		30
Aug.	9	9/9. 4 brood. 20 lb food				
Sep.	29	Fed syrup 10 lb		Net surplus		56
	1952					
	4					

It will be seen that the heading furnishes a complete clue to the identity of
the colony, which is No. 8 of the year 1951 and the queen was raised in 1949,
her number in 1950 being 2. On closing the records for the year a new sheet is
made out and its number put on the old one.

What is recorded on the cards is a matter of taste, but from what I have seen
in my travels, there is much vagueness. 'Strong and working well', 'gathering
pollen', and so on, do not lead anywhere, and as will be seen, my records
consist of things done to the colony, which have some bearing on its future.
Whenever I open a stock, I record its condition as seen. Thus 8/9 means that
there were 9 combs but the bees were only covering eight of them. When a
super is added, a note is made and likewise when one is removed. This enables me to see, not only how much each stock gathered, but when it did so.

Now for the last point. Where should these records be kept? Nine out of
ten will say 'in the hive', which is the place recommended by most guide
books, but this plan has no advantage to compensate for the disadvantages,
except the fact that it cannot be mislaid. In order to see the record, one must
open the hive, but surely the object of records is to know the state of the
colony *without* opening the hive. If we have numbered our stocks and our
records to match, we had far better keep the records indoors. We can then go
through them at any time and see exactly what needs attention. This indicates
that the records could be kept in a book, using a page for each stock, but I
think the looseleaf system is better. On a clipboard I keep a stout quarto card
and the record sheets are clipped on this in numerical order. Before visiting
the hives, I go through the sheets, take out those which indicate that the
colony needs attention and put them on top, returning them to proper order
when the work is done. A pencil tied to the clip is always handy. The clip
hangs in the honey house and is taken out when anything is to be done.

WEIGHING SYSTEM

For all ordinary purposes such records as this will suffice, but for the bee-
keeper who aims to get more exact information about the performance of his

charges, a system of weighing the hives is most illuminating. Not until one has practised weighing for a season, does one realize the peculiar manner in which the honey-flow fluctuates and discover that what seem to us favourable conditions are often quite the reverse. Weighing is almost the only way of testing the reaction of bees to certain circumstances, or the influence of various factors, like situation, variety of bee, type of hive and other things which more or less subtly affect their well-being.

There is also a practical side. Weighing enables us to time the application of supers or to feed at critical moments. Indeed, I am sure it will be agreed that if weighing were more generally and regularly practised, we should know much more about bees than we do.

Of course, it takes a good deal of time and trouble and the cost must be considered. Platform scales are expensive and not too portable, though they are the best for the purpose, because the hive does not need to be disturbed, but only one hive can be weighed, unless two people can help each other to lift.

For many years I have used a spring balance which accommodates 120 lb and shows half-pound variations. The hives I weigh are in a shed, and the balance is fixed to an iron rod 6 ft above the ground, so that it can be moved along to weigh any one of five hives. The floor of each hive has four strong eyelets, one at each corner screwed into the side. Through these a rope sling is threaded and this hangs on the hook of the balance. To facilitate lifting, the balance is attached to a pair of pulleys and a sash line runs through these for hauling. It takes very few minutes to weigh the whole five and record the weights on a card.

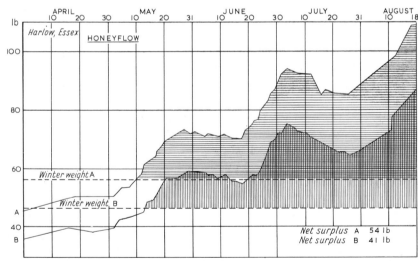

Normal honey-flow showing the June gap

In the active season at least one hive should be weighed daily and the figures correctly recorded. If the hive gains in weight beyond the capacity of the balance, surplus is taken off and weighed and this is allowed for in subsequent recordings. When put into graph form, the figures are very instructive, especially if accurate weather records are kept in the same district.

Little purpose would be served by weighing daily in winter. Once a week is then sufficient. At this time, a gentle and regular loss of weight indicates well-being. A sudden drop generally means robbing, while a stationary condition would imply that the bees were dead.

Part II

10
Whole or
Part Time?

In the foregoing chapters I have tried to cover the whole field of operations for the novice's first year of bee-keeping. By this time he will be pretty sure to have made up his mind whether he wants to continue. Much will depend on circumstances. He may have been unlucky in his first season, which may have been a poor one and his crop nil. He may have been so unfortunate as to lose a swarm. The bees may have been a difficult stock to handle and a series of severe stingings may have put him in fear of his charges. Any of these things, or all of them combined, may have disgusted him and made him decide to get rid of the bees to the first comer.

These things do happen. During the last war, large numbers of people started bee-keeping solely because, deprived of an unlimited supply of sugar, they hoped to make up for it by producing honey. That is a poor foundation for success in a craft which requires courage, patience and perseverance in no small degree. I have known a few keen bee-men who never got over their fear of bees, but they are the exception and unless one is prepared to handle bees confidently and be as much interested in their ways and welfare as their produce, a man will never be a good bee-keeper.

On the other hand, the novice may have struck a good-tempered stock of bees and an exceptionally good honey year. I have known more than one such take 100 lb of honey from his first hive, more than enough to repay his whole outlay. This is pretty good for three or four months' work, and may tempt anyone to launch out on a big scale. The next season may be a poor one, swarming may be troublesome and the previous season's profits used up.

If the beginner finds that his interest has grown, his next move must be determined by circumstances. If he intends to keep bees as a pleasant hobby, or has very little capital to spare, he will be wise to go slowly, getting one or two hives in readiness for new stocks he may hope to produce from the old one by methods I shall describe later. If he hopes to add to his income by keeping bees, or even contemplates making it his main occupation, he will perhaps be inclined to launch out boldly forthwith.

Opinions differ as to whether it is better to keep bees as part of a rural occupation, or to concentrate on them to the exclusion of other things. It is an important matter to decide and I will try and put both sides of the question. Those who favour whole-time bee-keeping contend that its busiest time is summer, when all other rural occupations are most active and this may lead to neglect of one or the other at critical moments. They also say that specialization tends to greater profit, because it gives more knowledge and experience.

I feel there are two weighty things to be said against this. The first is that old and well-tried advice against putting all one's eggs in one basket. Farming in this country has been badly up against it from time to time, but it rarely happens that all branches are in low water and while the mixed farmer may not make quite so much as the specialist, neither does he run such large risk of complete ruin. This is specially important because of the hazards of weather in such a climate as ours. A dry summer which favours bee-keeping is bad for vegetable growing and *vice versa* and I am sure that the prevalence of mixed farms in Britain is due to age-long experience of the dire disappointment which often awaits those who have only one string to the bow.

Secondly, it is undeniable that in matters connected with the land there is no such thing as water-tight compartments. Even market gardeners, who are as prone to specialize as anyone, find it worth while to keep a few cattle, sheep or pigs, not only because they provide an outlet for much material that would otherwise be wasted, but also because they permit of arable land being rested occasionally, without lying entirely idle.

BEE-KEEPING PLUS——?

Bee-keeping and fruit growing dovetail perfectly together and although it does not matter to the bee-keeper whether the fruit trees from which his bees get nectar belong to him or not, nor to the fruit grower that the fertilizing bees belong to someone else, it is clear that there is mutual profit and those who combine the two get it both ways. Admittedly, there is not the same close connection between bee-keeping and other rural pursuits, but that does not bar any mixing of other profitable enterprises with bee-keeping. You may keep poultry, for instance. The busiest time for this is winter and early spring, when birds are housed, and hatching and rearing proceeding. These are the dullest months for the bee-man. Those are two lines most attractive to men with a desire for active, self-reliant life in the country, but there are other things which appeal to various individuals who must decide for themselves, according to their knowledge and proclivities, to what extent they will combine with bee-keeping.

The chief objection to whole time bee-keeping is that it is not really practicable outside the choice areas of the country. Not more than fifty hives should be put in one spot and it will take at least two hundred to give a man full-time work. That means that he must spread his stocks over a wide area. In the best of conditions there must be considerable outlay on transport between the different centres. One must also remember that in these favoured spots, other bee-men are already established and there will be difficulty in securing good sites. The rule about over-population applies whether the bees belong to one man or several and the existence of apiaries on or near the sites of your choice, may force you to take much less favourable places.

FIND A MARKET

Another reason for exercising caution in expanding the apiary is the marketing problem. To the amateur this means nothing, but it is useless for the commercial bee-keeper to raise a crop of honey, if he cannot sell it, or only at a price below the cost of production. Taking them all round, the British people are not honey-minded, and the number of regular buyers is only a small proportion of the population. More than once in the past twenty years a heavy crop of honey has been difficult to dispose of and has had to be kept in stock till a poor season enabled it to be cleared.

The fluctuation of the seasons is the chief reason why the demand does not grow, and it is made worse because many small bee-keepers are over-anxious when they get a big crop to turn it into money and will often accept a price below production cost. This is very foolish, because properly harvested and packed honey keeps indefinitely, so that if we stick to a fair price and have a good quality left at the end of the year, customers can be kept supplied if the next season is a poor one.

By building up the apiary slowly, opportunity is given for a steady clientele to be secured. This is only possible if the customer knows he can get further supplies when he needs them.

HOW BIG ARE WE?

There is no doubt that most of British bee-keeping is on an amateur, a hobbyist basis. A very few hard-working individuals are on a full-time commercial basis but some of these have to supplement their supplies by importing honey in bulk, packing and selling it—suitably labelled as imported, of course. An even fewer number, mainly heather specialists, confine their activities to the production of comb honey.

The following figures (for 1975) give an indication of the position:

Estimated total number of colonies in England and Wales			176,448
Bee-keepers owning	1 to 10 colonies		28,108
,,	,,	11 to 39 ,,	2,878
,,	,,	40 to 100 ,,	317
,,	,,	over 100 ,,	115

(These figures are taken from official sources and are probably under-estimated.)

The vagaries of the British climate make it possible to get a honey surplus well in excess of 100 lb per colony in exceptionally good years, whereas in a bad year there may be no surplus at all. It is probably prudent to reckon that, taking a period of, say, ten years, good, bad and indifferent, a fair estimate of average surplus might be to the order of 30 lb per colony.

The Australian picture is:			
Estimated total number of colonies			524,000
Bee-keepers owning	5 to 99	colonies	4,487
,,	,,	100 to 199 ,,	487
,,	,,	200 to 299 ,,	271
,,	,,	300 to 399 ,,	172
,,	,,	400 to 499 ,,	107
,,	,,	500 to 599 ,,	72
,,	,,	600 and over ,,	162

The average yield was 85 lb per colony. Due to the vast area available and the range of climatic conditions, more bee-keeping is practised on a migratory system than in Britain. Total production in 1972 was 44,621 tons.

It is interesting to note that, in the year 1972/3, the Australian Honey Board was able to spend $A44,000 on salaries and administrative expenses and no less than $A81,000 on promotion, publicity and research.

It is estimated that there are over 4,000,000 colonies in the United States and that the average annual yield per colony is in the region of 58 lb. According to figures issued by the US Department of Agriculture, the total honey production for 1974 amounted to 185,338,000 lb, which increased to 190,000,000 lb in 1975. The estimated total US production for 1976 repeats the previous year's figure.

The following table indicates the number of colonies, yield per colony, and total production of honey and beeswax for the years 1960 to 1974.

Year	Number of colonies	Honey yield per colony	Production Honey	Beeswax
	Thousands	Pounds	1,000 pounds	1,000 pounds
1960	5.005	48.5	242.802	4.372
1961	4.992	51.3	255.868	4.720
1962	4.900	50.9	249.608	4.805
1963	4.849	55.0	266.778	4.828
1964	4.840	51.9	251.188	4.672
1965	4.718	51.3	241.849	4.697
1966	4.646	52.0	241.576	4.615
1967	4.635	46.6	215.780	4.386
1968	4.539	42.2	191.391	3.797
1969	4.433	60.3	267.485	5.171
1970	4.290	51.7	221.842	4.377
1971	4.110	48.0	197.428	3.585
1972	4.067	52.6	213.959	3.986
1973	4.103	57.9	237.657	4.231
1974	4.194	44.2	185.338	3.413

BEE-KEEPING IN BRITAIN

So many enthusiasts have, in the modern renaissance of bee-keeping, written in advocacy of its adoption as a means of livelihood, that I feel it to be urgent to sound a warning note and point out the snags.

For those who intend to adopt it, locality is of vital importance. Not only is the normal crop in a good district as much as three times that in a poor one, but the quality of honey is more generally favoured and marketable. Personally, I like strong dark honey, but it is not the popular taste, which inclines to the mild, more attractive-looking product of clover. It is a significant fact that the winners of the W.B.C. Gold Medal, the Blue Ribbon of Bee-keeping, have, with perhaps two exceptions, been located on a chalk or limestone soil. Not one resides in Norfolk or Lincoln, counties which produce abundant honey, but apparently not of the supreme quality.

For quality plus quantity, I should put the Cotswolds in first place, the Chilterns next and then the South Downs, but almost any place where the chalk comes to the surface is good bee country. Sometimes such spots are found in the middle of a general desert, so to speak, of clay, as in the district round Saffron Walden in Essex, where light clover honey is abundantly produced though it is rarely obtainable a few miles away.

On the light alluvial soils of the fens, a good account is always rendered by bees and these have an advantage over the west, because they are drier; in wet seasons they always outstrip the Cotswolds in production.

While the richest soils produce honey in greatest abundance, the poorest have compensations. Whole areas scattered about the country, but chiefly in the hills of Yorkshire, Northumberland, Durham, North Wales and the Scottish Highlands, are given up to the various species of heather. Here again, production varies with the nature of the subsoil. On some, the annual crop is, weather permitting, good. On others it is uncertain and fluctuating, but wherever it is available in sufficient quantity, heather is a valuable honey plant and those within reasonable reach of it usually arrange to put at least some of their stock on the moors while it is in bloom. Happy the bee-man who lives in a good clover district and has within a few miles a heathy waste to which he can take his bees in August.

Generally speaking, clay soils are not favourable for bee-farming on a large scale. On such land bees often make a good show in the early part of the year, because trees are generally abundant, but later production is often indifferent, because of the clover's need for exceptional warmth.

This limitation of locality choice is enough, apart from other considerations, to show that whole-time bee-keeping can only be for a few, and they are, indeed, few. Even among these, not many seem able to carry on without straying from the path of pure honey production. They are almost all queen breeders or appliance dealers and it must be obvious that while there is unbounded scope for honey production, queen breeding and appliance dealing can only expand within narrow limits.

It is the fact that more than a hundred hives in one spot is economically unprofitable in the normal district which convinces me that bee-keeping should not be more than a part-time occupation. Even in the active season, less than that number will not occupy all a man's time, while, however many he has, he will have little to do in winter. Is it not better, therefore, to reduce the number to that which shows the highest relative return and marry it to some occupation of a congenial nature?

Except for the work of extracting and bottling honey, everything has to be done by hand and in an exceptionally good season it is not easy to find people ready to help with the harvest.

In Britain the Bee Farmers' Association fixes a minimum of forty hives as qualification for membership, and I suggest this figure as suitable for those who keep bees as part only of the activities of a smallholding. This number will probably give the best return for the outlay on efficient plant.

11
Propagating Bees

Whether the bee-keeper whose first season fills him with enthusiasm, merely desires to continue it as an interesting and profitable hobby, or aims to make it all or part of his breadwinning activity, he will certainly want to increase the number of his colonies. If he is so rash as to plunge into large-scale work at once, he will have to buy bees, but apart from the fact that, in my experience, it takes at least three years to acquire sufficient competence to manage a large apiary, it is by far the most costly and risky method.

At the present time, not only are bees expensive, but are so much in demand that to get any considerable number of stocks, they would have to be obtained from varied sources and the inexperienced man may have very inferior, if not actually diseased, bees wished on him. On no account should he buy from strangers or any but the most reputable dealers. It is far better to exercise restraint and build up the apiary from present stock, acquiring at the same time more knowledge and skill in management. If it is desired to have a special strain or race, one or two stocks can be obtained from a reliable source and can be used to make increase to a modest extent. In this way, at a cost not much in excess of the hives and frames of foundation, a good-sized apiary can be built up in two or three seasons.

Like other forms of livestock, bees are prone to increase and multiply. Indeed, the problem often is to curtail this propensity, for the ultimate aim of bee-keeping is not more bees, but more honey. If this end is to be attained, the increase of bees must be controlled and regulated, for it is useless to increase the number of colonies if it results, as it well may do, in a reduced quantity of honey.

Before the introduction of frame hives there was only one way to increase one's stock of bees—natural swarming—and I have explained how the old time bee-keeper did, to some extent, regulate the natural increase. Nowadays several different methods of propagating bees may be followed. The analogy I drew between a bee colony and a tree can be extended to the means of propagation, for just as one plant can be made into several by cutting or dividing it up into suitable parts, so two or more stocks of bees can be made from one by a process of division, a method quite out of the question with any other form of livestock.

Before making any attempt to increase stocks, empty hives and frames of foundation must be in readiness and as the demand for bee-keeping material of all kinds is likely to outrun supply during the active season, it is well to look ahead and get what is likely to be required during the winter.

INCREASING STOCKS

If the bee-keeper's ambitions are modest and his desire is to have two stocks instead of one, I do not think he can do better than let his stock swarm and proceed by the Pagden method described on page 57. Providing he makes sure that both lots are well nourished by feeding them if there is not a good honey-flow, he may reasonably expect to get not only the new stock but a reasonable amount of surplus honey. It must be remembered that the swarm has an old queen. He may know the year of her birth or he may not, but if, as is often the case, he started with a swarm the previous year, she may be three years old or more, past her best and perhaps unable to sustain the colony through the coming winter. He will thus be pretty much where he started, except that his new stock will have a queen at her best period. It is wise, therefore, to make a small nucleus, using one of the queen cells which would otherwise have been destroyed. This nucleus must stand beside the swarm and, after the honey-flow, be united to it, after the old queen has been removed.

It may happen that the stock does not swarm. If the queen is in her second season, the odds are against it and in that case the procedure is rather different. Two courses may be followed: one or more queens may be bought from those who specialize in breeding them, and given to nuclei made from the stock on hand; or one may be raised in the stock.

MAKING A NUCLEUS

Although making a nucleus is a fairly simple matter—I have already stated how it is done in Chapter 4—to do so without materially reducing the strength of the stock requires a little care. The bees we need in the stock are chiefly the mature foragers. At this time, brood-raising has usually passed its peak and the younger newly emerged bees which act as nurses can very well be spared. These are, in fact, the ones most suitable for forming nuclei; first, because not having flown out, they do not leave the nucleus and second, because they accept a new queen much more readily than older bees. Mr Snelgrove gives a method by which a nucleus may be made up of such young bees most effectively and it has the merit that it does not entail searching for the queen, a job which even experienced men find difficult at times.

Three combs, one at least containing plenty of unsealed brood, and the others a good supply of honey and pollen, are taken out of the hive and the bees shaken back. The combs are put in a brood-chamber, with the brood in the centre. A queen excluder is put over the stock brood-chamber, the three-comb lot stood on this and covered over. In a very short time young bees will come up through the excluder and cover the three combs. In less than half an hour it can be removed and put on the place prepared for it.

INTRODUCING A QUEEN

The new queen—which will have been ordered in advance—will arrive in an introducing cage, with instructions for introducing her, but such a nucleus will usually accept any kind of queen twenty-four hours after it has been formed, without any special precautions. I shall say something about methods of introducing queens later, for it is not always so easy as in this instance.

If a queen is not obtained from outside, one must be raised. A nucleus is taken out of the stock, but this time it must contain the queen and, so long as the old colony has eggs, it may be trusted to raise a queen. If inspected after ten days, the cells will be found well advanced and all but one can be cut out.

Under reasonably good conditions, the stock should not be much handicapped by making this nucleus, but store honey almost as freely as if nothing had happened.

PLANNED QUEEN BREEDING

Certain disadvantages attend the rough and ready method of raising a queen just described and to make clear what they are I must go a little more closely into the subject of queen rearing. Queens are naturally reared in three distinct ways. The most usual is the swarming type, made when the bees deliberately prepare to divide into two or more parts. Swarming queen cells are always the most numerous: I have rarely found less than eight in a swarmed stock and some races of bees make an endless number—indeed as many as a hundred have been found on one comb. Because of the fact that swarming results from a physiological urge in the colony, swarm cells are well provided with royal jelly and the queens produced have the best development of which their stock is capable.

The second type of queen cell is the supersedure cell. When a queen is producing few eggs, either because she is of poor quality, or too old, the workers often supersede her. They may also do so, in spite of good egg-laying, because she has been damaged by the loss of a leg or wing. In supersedure, the general rule that two queens will not live together in a hive is sometimes broken. When the bees decide to supersede, they make a few queen cells—sometimes only one and seldom more than four or five. Whereas swarming cells are usually on the edges of the combs, supersedure cells are built out from the comb face. These cells are generally long and pointed and the queens raised from them are often of exceptionally good quality. Even after the young queen has emerged, the old one often remains until after she has mated and begun to lay. Mr Manley mentions one which remained with her daughter more than a year. More often, I think, the old queen is disposed of when the young one emerges and sometimes the old one will lead out a swarm. This will often return to the hive, having, in the meantime, got rid of the old queen. There seems to be no regular rule in the procedure and because it frequently takes place unknown to the bee-keeper, we do not know enough about it to form any definite conclusions.

SUPERSEDURE QUEENS

Many bee-keepers consider that supersedure queens are the best and there is no doubt that they are well nourished in infancy, but colonies which constantly renew their queens in this way are often influenced by their inferiority. Only last year one of my stocks superseded its queen twice, first in the spring and again in autumn, but never managed to build up more than five combs of brood. The chief claim to superiority made for supersedure queens is that they are a step in the direction of the elimination of the swarming impulse and this is certainly a very important matter from the bee-keeper's standpoint.

EMERGENCY QUEENS

The third type of queen cell is that raised in a hurry because the queen has been lost. She may have died from old age before the bees thought of superseding her—it would seem that some strains never use this method—or she may have been killed, or left out of the hive during manipulation. Queen cells built in these circumstances are very variable, both in number and quality. In a strong stock, a good many are sometimes raised, but a weak one only makes a few. Owing to the fact that in this emergency the bees often make use of three-day-old larvæ, emergency queens are often of poor quality. The cells made under these conditions are generally in the middle of the comb and are short and stumpy.

RAISING QUEENS

The above facts present us with a dilemma. On the one hand, we do not want to encourage the swarming habit. I doubt whether it would be an advantage to eliminate it altogether, but it is such a nuisance that we must try and keep it to a minimum. Against this we have the fact that swarming cells produce the largest number and best quality queens.

Out of this difficulty there have evolved techniques of queen rearing so varied that there is almost a library of books dealing solely with this fascinating topic. In one of the most comprehensive, L. E. Snelgrove's *Queen Rearing*, no less than twenty-four different ways of raising queens are described. It would be unnecessary for me to give such a large number, but I will try and make clear the principles which should guide the queen rearer in his work.

1 Queens should be raised from the best stock.
2 They should be from larvæ not more than 36 hours' old.
3 They should be fed lavishly with royal jelly.
4 They must be raised when drones are available.

It is clear that if we have a number of stocks of varying quality, it is quite likely that the poorest may be the first to swarm. These will raise good queen cells, but they will not be likely to give better quality colonies.

The best stock will, perhaps, not swarm at all and if we remove its queen to cause it to form queen cells, these may be raised from older larvæ and be inadequately nourished.

The remedy for this is quite simple, because we can at any time take eggs from one colony and give them to another. By taking them from the best stock and giving them to one about to swarm, we satisfy both requirements. Some not unimportant minor details need fulfilment and these I will deal with in turn.

SECURING EGGS

To secure a supply of eggs from the selected stock is the first step and the plan almost invariably followed is to insert a frame of foundation in the centre of the brood-nest, a comb being removed to make room for it. An empty one, or one with honey only, should be taken from outside and the others pushed back to leave room in the centre. The foundation in this frame should not be wired, but securely fixed in the top bar. Some prefer to use a shallow frame, which is more easily handled in subsequent operations. On the fifth or sixth day after this, the frame should be inspected and will normally have been built into comb and more or less filled with eggs; those near the top and centre may, indeed, have hatched.

If this comb is put into a stock which is already building queen cells, it will almost certainly raise queens on the comb we have given them, especially if we destroy the cells they have already made, but if left to chance they may only raise one or two. Moreover it will have been noticed that naturally raised queen cells built under the swarming impulse are usually at the bottom of the comb, or in some hollow space where there is plenty of room for the cell to extend downwards and it is a strong inducement to cell construction to arrange that eggs do, in fact, have plenty of space below them. The simplest way to do this, is to cut away the comb below a long row of eggs, so as to leave at least two inches of clear space. More eggs can be thus exposed by cutting the comb in zigzag fashion, as originally recommended by Dr Miller. His plan was to insert triangular pieces of foundation, but in a strong stock the empty space may very quickly be filled with drone comb which has, in any case, to be cut out.

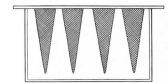

Strips of foundation for mass queen rearing

Another thing which happens in natural cell raising is the junction of two cells together, larvæ in adjacent cells being fed as queens. These are hard to remove without damage and it is therefore an advantage to use only one in three cells, destroying the eggs in the two cells between those selected and slightly enlarging the mouth of each selected cell by carefully twirling a match-stick round the edge.

Even more room for queen cells can be provided by another method of dealing with the comb of eggs. Instead of cutting it to allow space below the cells, it is laid flat over the top of a pre-swarming stock, supported in a wooden frame, so that all the cells on one side have a clear space below them. As before, it is desirable to prepare the comb by destroying all the eggs in alternate rows and two out of three in the rows selected.

THE ALLEY METHOD

A third method, introduced by Alley (USA), is to cut the comb of eggs in strips, so that a row of cells with eggs runs along the middle. This strip is attached to the underside of the top bar of a shallow frame, by painting melted beeswax along the bar and pressing the strip onto it, afterwards making it more secure by painting more wax along each side of the strip. Plenty of wax should be used by putting on two or three coats before fixing the strip. Again, two out of three cells are destroyed, so that the rest are well spaced apart.

This leaves about ten to twelve cells on the bar, and two bars can be used by fitting a second one into the middle of the frame by tongue and groove at each end.

Ten days after the cells prepared by one of these methods are given to the pre-swarming stock, they will be sealed over and ready for removal to nuclei.

The use of pre-swarming is not always possible, and restricts the choice of time for raising queen-cells, so that although ideal, it is seldom used in prac-tice. Fortunately there are other ways of inducing stocks not naturally dis-posed to build cells to do so. Many years ago, the late G. M. Doolittle, a

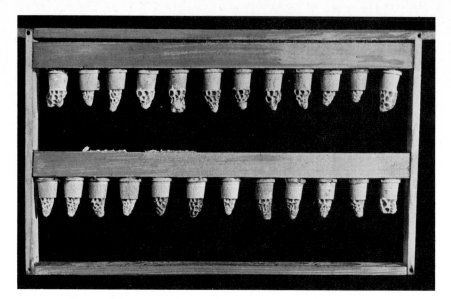

The Alley method of queen rearing involves the formation of bars of artificial queen cells

pioneer of planned queen rearing, pointed out that if a comb of brood is isolated from the main body, the bees in attendance on it will start queen cells, under the impression that their queen is failing. This happens frequently when a brood-chamber is raised above the supers in the Demaree plan.

A stock occupying two brood-chambers is the best to use and most of the unsealed brood should be arranged in the top chamber, so that there are plenty of nurse bees in attendance. If the Snelgrove board is employed, it should not be put on till a day after the top chamber is raised, so that nurse bees can come up. After that it is a help to put the Snelgrove board underneath, leaving one upper exit open. This should not be closed or the lower one opened until the queen cells have been sealed over. There should be plenty of food in the upper box.

This plan allows us to use any strong stock in the apiary and in fact the same stock which provides the selected eggs can be used, though, if we wish to raise a good number of cells conveniently placed for removal, it is better to proceed in exactly the same way.

The advantage of rearing queens over a strong colony in this way is that the queens are well nourished, owing to the ample supply of young bees to furnish royal jelly, but sometimes there is difficulty in getting cells started in such a stock. They will rarely fail to make one or two but, as in natural supersedure, they sometimes content themselves with a single cell. If we contemplate substantial increase or wish to raise queens for sale, this would be a waste of time and trouble. The ideal is to try and get about twenty cells completed in a queen-raising colony.

GETTING CELLS STARTED

The difficulty is to get the cells started. Once they have been started any bees will, in accordance with their invariable custom, carry on and complete the work.

The only kind of colony we can be sure will start queen cells is one without queen or brood, but as such a colony cannot be relied on to provide sufficient nourishment to the young queens, it is used only to *start* the cells, which are then given to a queenright stock prepared as above to complete.

The bees to be used for starting cells are taken from a strong well-fed stock and put, with two combs or more of honey and pollen, but *no brood*, into a closed box. A W.B.C. brood-box, with a sheet of perforated zinc nailed on its underside, raised on battens to allow ample ventilation, will serve. The combs should be arranged so that there is space in the middle for another—the frame or bar of prepared eggs.

COVERING

The covering can be of cloth or wooden, but it must be impossible for the bees to escape. Select the combs of food, making quite sure there are no eggs or larvæ in them and put them in the box. Having found the queen of the stock and put her temporarily out of the way, shake the bees from five combs into the box, close it up immediately and take it to a cool dark place—a cellar is ideal. In about three to seven hours' time, the confined bees will have elaborated plenty of brood food. The frame of eggs is now put in the space left for it. To prevent the bees escaping, use a carbolic cloth, drawing it over as the cover is removed. Quickly insert the frame and cover up again. A feeder containing thin syrup should be placed over a feed-hole in the cover and the box left undisturbed for twenty-four hours. By this time all queen cells the bees are likely to make will have been started and the comb should be removed and given to the strong stock which has been prepared to complete the cells. The box of bees can then be taken back to its hive and the bees shaken in.

COMMERCIALLY RAISED QUEENS

Commercial queen breeders usually raise queens in artificial cells. Frames, which vary somewhat in pattern, are fitted with wooden cups so that they can be readily removed. To each of these cups is attached a cell made of beeswax by dipping a shaped 'forming stick' into the melted wax several times until a thin cup, shaped like a natural queen cell, has been formed. Into each of these cups a drop of royal jelly obtained from a queen cell is placed and a very young larva, not more than thirty-six hours old, is transferred to it by means of a 'grafting tool', variously made from a quill, a wooden match stick, or a metal needle sold for the purpose. This is a delicate operation, for great care must be taken not to injure the larva or drown it in a superabundance of jelly. It is indeed a troublesome operation, scarcely to be undertaken by the amateur, and even professional queen breeders, notably the famous American, Jay Smith, have abandoned it in favour of Alley's method. Its one merit is that the bees are more ready to accept a cell formed like a queen cup and furnished with jelly and larva, so that it is not necessary to use a cell-starting stock.

The necessity of having drones on hand to fertilize the queens restricts the queen-rearing season to three or four summer months.

MATING QUEENS

Having secured a number of queen cells, we must make arrangements to accommodate the young virgins as soon as they mature. If there are a number of queen cells in a stock, the first virgin to emerge will usually destroy the others in a short time. It does not invariably happen, for the workers sometimes form a guard over unopened cells and prevent the queen from reaching them. She will then lead out a swarm or cast and another virgin will take her place. Quite often you may find that after a swarm has issued, especially if it happens after a wet spell, several virgin queens are at large in the stock. This is sometimes the reason for a swarm leaving again, after it has been returned to the old stock by the Demaree method.

The queen normally leaves her cell on the sixteenth day after the egg was laid and takes her first flight not less than six days after emergence. This flight is only a reconnaissance. Like all other bees leaving the hive for the first time, she hovers face to the entrance, marking its appearance and environs and gradually extends her circles until the spot is impressed on her memory. Sooner or later she takes the extended flight during which she meets the drone, after which she returns to the hive with his organs adhering to her. These are removed with the assistance of the workers and the mating is complete. At one time it was supposed that a queen never mated more than once, but plenty of evidence has accumulated to show that queens usually mate a number of times, frequently up to seven or even nine times, presumably because the first unions were ineffective. It has been fairly well established that, unless fertilized within four weeks, she is incapable of it. In practice, three weeks seems to be the limit and the average works out at about ten days.

To ensure successful mating, the presence of drone bees in the vicinity is an essential factor.

This requirement imposes a strict limit on the time of year during which queens can be reared. Drones rarely appear on the wing before the middle of April and, in most places, are destroyed by the workers in August, though a queenless colony will suffer its drones to live through the winter. It seems doubtful that they would fly or be effective late in autumn. From about the end of May until the end of July is the most favourable time for mating.

IMPORTANCE OF THE DRONE

Since the quality of a stock of bees depends on both parents, the drone is an important factor in the propagation of bees with desirable traits, but it is, unfortunately, a difficult matter to control. Mating cannot, so far as we know, be naturally accomplished except on the wing, so that we are obliged to permit our virgins to fly at will and cannot choose the partner with precision, as with other livestock. This has always been a serious obstacle to the improvement of stocks of bees and, as I have pointed out, makes it almost impossible to maintain a pure race of bees without continual importation of queens.

Queen-breeding establishments are, as far as possible, set up in districts remote from other apiaries, but even these are by no means secure. Wild stocks often exist in unsuspected places and there have been instances of mismating by queens that were raised on an island several miles from the mainland. Allowing a queen a possible flight range of two miles and a drone the same or even more, four miles cannot be considered a safe distance between the queen's hive and that of the drone.

In spite of this, queen breeders, even in not unduly isolated places, generally manage to procure pure mating for most of their queens by first controlling the production of drones in their own apiaries. A stock carrying the desired qualities is specially encouraged to raise drones by giving it drone comb, which queens, for some unexplained reason, are always ready to fill in spring and early summer. Stocks not possessing the desired qualities are deprived of drones by destroying drone comb and trapping any drones which, nevertheless, reach maturity. This is done by attaching before the hive entrance a chamber, with excluder zinc, through which drones cannot pass, as its only exit. A cone-shaped entrance leads from the hive into this chamber, so that drones entering it can neither return to the hive nor reach the open.

I know one queen breeder who, to reduce the risk of poor mating, distributes high-quality queens among the other bee-keepers in his district. By these means, the balance of probability in favour of correct mating is more or less secured.

ARTIFICIAL INSEMINATION

About 1927, L. R. Watson (USA) devised an instrument for the artificial insemination of queens. This and the technique for using it have been greatly improved, so that it is now possible to control mating by this means. So far, artificially mated queens have not proved as fully satisfactory layers as naturally mated queens, but the process is clearly capable of helping greatly to improve our stock. The apparatus is costly and the work highly skilled, so that its use will probably be confined to special queen-mating stations.

MAKING NUCLEUS HIVES

The provision of bees for the young queen is always a problem for the owner of only a few colonies who wishes to increase his stocks substantially, for he can only do so at the expense of his honey crop and, unless moderate in his aims or exceptionally skilful, he can easily ruin his stocks and still fail to obtain the number of mated queens he desires.

I have already described the methods and precautions to be observed when making nuclei. On the whole, it is most satisfactory to make only one from each strong stock, as this can be done without seriously affecting its prosperity. A poor colony, unlikely to build up to profit-making capacity, can, on the other hand, be de-queened and broken up to make two or more nuclei. If the aim is to make increase and sacrifice the potential honey crop, even a strong stock can be broken up into three or, with due precautions, five or even more.

Generally speaking, it is desirable that a nucleus should have three combs—one of brood and two well stored with food—but if well covered with bees, two are quite satisfactory. An abundance of food is essential to a nucleus, not only for the nourishment of its members, who are not able to forage adequately for themselves, but because the acceptance of queen cells or queens is much likelier if there is plenty of food. Nucleus hives can be purchased or made, either to hold a single nucleus up to four combs, or two or more separate nuclei separated by partitions, with separate covers and entrances on opposite ends or sides. An ordinary brood-chamber can thus be divided into four parts, each of which will hold a two-comb nucleus. Those at the side have an entrance made in the side wall, the two middle ones have theirs respectively in the front and back walls. The advantage of this is that, while quite separate from each other, they combine to maintain warmth.

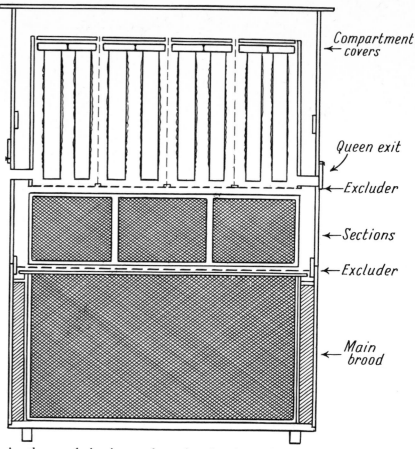

Compartment
covers

Queen exit

Excluder

Sections

Excluder

Main
brood

Queen-rearing hive

Another good plan is to make such a chamber to fit over a queenright hive. The bottom is covered with wire cloth or perforated zinc and the partitions made to slide in grooves. It is essential, in either case, that bees cannot get from one compartment to another and each must have a separate cover, so that only one need be exposed at a time. In such a chamber, the whole process of queen raising and mating can be accomplished, by leaving out the partitions until the cells are sealed and one distributed to each nucleus. If there are not enough bees to furnish all the nuclei, one or more can be made up from other stocks as described.

Nuclei made up from the bees which have been raising queens, need only to be deprived of all but one sealed cell which, in due time, should produce a virgin queen without further attention, but certain precautions must be observed when introducing queen cells to strange nuclei. At least two days should elapse before a newly formed nucleus is given a strange queen cell. If it consists, as it should, of young bees it will usually accept the cell, but sometimes it will be treated as an intruder and torn down. This subject will be treated later. The cell can be given at any time after sealing, but it is usual to wait till it is almost time for the queen to emerge. It is seven days after sealing before she is ready to do so and, towards the end of that time, the bees remove the wax from the tip of the cell, leaving only the tough cocoon spun by the larva. In this state a cell is said to be 'ripe' and within hours the virgin will make her exit by cutting a circular piece out of the cap.

Great care must be taken in removing a cell. The comb should never be shaken, but the bees brushed or smoked off it. A good margin, not less than half an inch wide, should be allowed and to be on the safe side, rather more left on top. Cells affixed to a bar are removed by slipping a knife under the wax used to secure them.

The cell should be put in at the top of the nucleus, the simplest plan being to lower it between two combs and put a tack through the piece of comb at the cell's base, pushing the tack firmly into the frame.

All operations connected with queen rearing should be carried out as quickly as possible, avoiding exposure of larvæ or queen cells to cold wind or bright sun. During warm days there is not much risk, but if it is necessary to transfer cells on a cold day, each cell must be protected. A good plan is to put them all in a small box and keep this in the trouser pocket, taking each cell out as the nucleus is opened to receive it.

QUEEN CELL CAGES

If a number of queen cells have arrived at maturity and there are not enough nuclei to receive them, they can be kept for a few days in nursery cages. There are various forms of these cages. The kind generally sold by appliance dealers is made out of a block of wood $2\frac{1}{2}$ in square and 1 in thick. A hole is bored right through it and covered over with wire cloth. From the top, two other holes are bored through to the central hole. One of these is filled with candy, made by mixing honey with finely powdered sugar to the consistency of dough. The queen cell is inserted in the other hole, so that its point hangs down inside. The candy serves to nourish the queen while she is caged and the cage can be given to the colony to be requeened, by laying it over the frame tops. More candy should be put in if much has been used and in due time the bees will eat their way in and release the queen.

A cage can be made at home very simply out of a couple of wine bottle corks and a piece of wire cloth. The cloth is made into a tube fitting closely round the corks, one of which is hollowed out inside and stuffed with candy. The other has the queen cell stuck to its underside with wax. These cages must be kept in a hive while occupied and a number of them can be secured into a frame. The commercial form is sold in sets of eighteen fitted into a special frame.

QUEEN INTRODUCTION

The closeness with which bees stick to members of their own hive is only equalled by the ferocity with which they will attack strange bees. It is believed by many that bees recognize each other by scent, and although some experienced men dispute it, lots of things seem to confirm it. Our knowledge of the senses of bees is far from complete, and it is a pity to be dogmatic on such points which are very difficult to determine. However it may be, strange bees are always attacked if they try to enter a hive, unless they are either very young, or are loaded with honey, and whether it is due to strange odour, or a consciousness of potential danger in the stranger, seems purely an academic matter.

At any rate, this antipathy to strangers is intensified in the case of queens, for not only will a strange queen not be allowed in a hive which already has one, but even after its queen has been lost a stock will not accept another unless certain conditions are fulfilled.

The subject of queen introduction is one of great complexity and wide divergence of opinion. Methods which one claims to have found always successful will be decried by another, in whose hands they have failed. A plan quite satisfactory for giving a queen to a nucleus may be useless for requeening a strong stock, and a stock which could be requeened at one time of year with scarcely any precautions, will at another time refuse a new queen in any circumstances. No known method commands one hundred per cent success but if the basic principles are understood and suitable methods applied in each case, the percentage of failure should not be greater than with any other form of propagation. No gardener, however skilled, expects every one of a batch of cuttings to root.

REASONS FOR INTRODUCTION

The circumstances under which queens have to be introduced are varied. The most common is the formation of nuclei for increase of stocks and this rarely presents difficulty, mainly because the bees are young and not aggressive. Queenlessness from death is perhaps the next most common occasion; improvement of stock by introducing a queen of better pedigree comes next; and last, most difficult of all, is the requeening of a vicious stock, tiresome and dangerous to handle.

There are innumerable ways of introducing queens and it would be beyond the scope of this book to deal with them all. I can only describe the more generally practised methods.

Beyond question, the easiest of all is to make a nucleus colony and give the queen to it. If a fertile queen is given to such a nucleus as soon as it is seen to be looking for a queen, she will be received with joy. Newly hatched virgins will also be accepted by such bees, but they do sometimes tear down queen cells and kill the occupant, so that it is better to give a queen cell in one of the tubular cages.

Once having safely introduced a queen to a nucleus, the way is made much easier to requeen a queenless stock, for as soon as the queen has settled down and begun to lay, the nucleus can be united to the queenless stock with as nearly absolute certainty as could be wished.

It may be, of course, that one is not able to form a nucleus, as when a stock has been found without queen or brood: all the bees are past the house stage and the queen must then be introduced directly.

ONE QUEEN ONLY

The first and most essential thing is to make sure there is no queen in the stock when the new one is introduced. I am sure that most cases of failure by beginners are due to the fact that they jumped to the conclusion that the stock had no queen when in fact it had.

If, in the height of the breeding season, a thorough search fails to reveal either queen or eggs, there is little doubt that the colony has swarmed so it may be a month before the new queen begins to lay: she is young and active and the novice may not succeed in finding her, as she will dodge from comb to comb as each is exposed to the light. It is necessary, when looking for queens, to cause as little agitation as possible. To keep pouring smoke onto the combs, drives the bees hither and thither and makes it very hard to find a queen running with them. During a good honey-flow it is best to do without smoke, open the hive quietly and take out each comb smoothly and carefully. At other times one or two good puffs of smoke should be blown in and the hive left alone for a couple of minutes, so that the bees may gorge. They will then stay quietly on the combs.

A fairly reliable guide is the general behaviour of the bees. If they are clustered closely, especially towards the centre, there is good reason to think there is a queen and this can be confirmed by closely inspecting the cells in the centre of the middle combs. If these are empty and highly polished over a small area, there is almost certainly a queen on the point of laying. The hive should be closed up and inspected a week later.

If, on the other hand, the bees are loosely scattered about the comb and sharply resent interference, there is every reason to suspect queenlessness. If a comb of eggs is available, it can be put in the centre of the brood-nest and left for a week. If no queen cells are started, it is safe to conclude there is a queen. If queen cells have been begun, the stock is queenless. It may be allowed to raise its own queen if no other is available, otherwise the cells should be destroyed and a new queen introduced.

INTRODUCING CAGE

When it is desired to requeen a stock with a superior race or strain, the old queen must be found and removed. The new queen must be ordered in advance and will come in a special travelling cage, accompanied by instructions for using it. The cage almost universally employed for this is cut out of a solid block of wood, three holes are bored in it big enough to accommodate the queen and her attendants, who occupy two compartments, while the third is filled with candy, which serves for food on the journey and as a medium of introduction. All three chambers communicate with each other and are covered with wire cloth. At each end of the block, holes are bored to communicate with the centre; that at one end is covered with wire put on after the bees were inserted; the hole at the other end is stuffed with candy, covered with a piece of paper.

When introducing a queen alone, it is essential that the bees to which she is being introduced should be able to feed her and groom her to some extent. The size of gauze to allow this has to be $\frac{1}{8}$ in mesh, while that used commonly on travelling cages is too fine. It is in fact undesirable for these cages to be used for introduction because they may be contaminated. The type of cage recommended by Dr Butler of Rothamsted is preferable for the actual introduction and these cages are available cheaply from the appliance manufacturers.

It is suggested that the following method is preferable to the instructions usually given for the travelling cage. When it is received, the travelling cage containing the queen and her attendant workers should be placed on its side close to the open feed-hole of the colony to which the queen is to be introduced. Do not remove the plugs from the ends of the cage, but leave the cage exposed to the bees for twenty-four to forty-eight hours. This 'conditioning' greatly improves the chances of successful introduction, especially of queens which have been caged for a week or more, as may happen with imported queens.

After the conditioning period the travelling cage should be taken indoors and it will then be necessary to separate the queen from the attendant workers and put her into the introducing cage immediately before this is put into the hive. Perhaps the simplest way is to open the cage against a closed window: the queen will run onto the glass and can then either be picked up and put into the introducing cage or gently coaxed into it. When the queen is in the introducing cage the open end should be covered with a single thickness of newspaper held on with a rubber band. The cage with the queen—containing no bees or food—should be put into the hive between two brood-combs well covered with bees within half an hour.

FINDING THE QUEEN

To find the queen is the most tiresome task when a vicious stock is to be dealt with. One way of doing it is to remove the stock a few yards from its stand. This can be done at night, the entrance being stuffed up with grass or rags for safety's sake. An empty hive containing combs with food, but *no brood*, is put on the old stand and the new queen in her cage is laid over the centre combs. Next day, the bees in the old stock can be released; numbers will soon return to their old stand and stay with the queen. A day or so later, the old stock, which will now contain only young inoffensive bees, can be inspected and the old queen killed. This lot can then be rejoined to the other by the newspaper method.

Bees without combs, brood or queen, will almost invariably accept a queen, fertile or virgin, during the honey season, especially if they are confined in a box or hive. A queenless stock may thus be quickly requeened by shaking three or four combs of bees into a box, shutting it up for four or five hours— allowing suitable ventilation—and then dropping the queen among them. They should be fed until the next day and then shaken out in front of the hive they came from. They and their new queen will enter peacefully.

THE LAYING WORKER

One of the most difficult problems connected with requeening is the laying worker. After a stock has been queenless—perhaps unknown to the bee-keeper —it often develops these pests. It is generally believed that they are workers which, in the course of their upbringing, have received some of the food appropriate to queen larvæ, causing their ovaries to be partly developed, so that they can lay a few eggs. Whether or not they do so when a queen is present is doubtful, but in a queenless stock they frequently do. Their brood can generally be recognized by the irregular manner of its distribution. A healthy queen puts her eggs in regular spirals, so that rows of eggs are followed by rows of larvæ in succession. Laying workers put their eggs haphazard, in a cell here and there with gaps between. It has been shown that several are often present, but as they do not differ in appearance from normal workers, they cannot be detected. If not recognizable as eggs, their brood is known later because only drones result, though reared in worker cells. An old worn-out queen also produces this kind of brood, but still lays in good order and she can be found and removed.

As a stock with laying workers will not accept a queen in any form, nor raise one from eggs given to it, the best thing to do is to unite it to a strong one by the paper method.

12
Honey in
the Comb

I deliberately refrained from giving details about the production of comb honey in the first part of this book because it is, in my opinion, a mistake for beginners to attempt what is admitted to be the most difficult thing to accomplish satisfactorily. Indeed, some of the biggest commercial bee-men in Britain will have nothing to do with this side of the craft, for several reasons. It is the hardest thing in the world to produce it, without at the same time, forcing the bees to swarm. Even at its best, it results in as much as twenty-five per cent less crop and it is necessary, if a good crop has been secured, to sell it quickly or granulation may make it unsaleable.

The older writers quite innocently advised the beginner to super with sections because, until extracting apparatus became popular, few were able to do other than cut out comb and strain it through a muslin bag. Sections avoided that, as well as the cost of extracting plant, but it was always accompanied by troublesome swarming and in all but the very best seasons, the crop was small and the sections of poor quality.

To the connoisseur, however, comb honey is esteemed far beyond any other form and sections have always been saleable at a higher price than that of honey in jars, even when heavy imports of honey have spoilt the market for the extracted product.

In order to enjoy comb honey it is not at all necessary to raise it in sections. Ordinary shallow combs, if fitted with thin unwired foundation, can be stored away as they are and slices cut out for the table as required. Heather honey, which is jelly-like and does not run out of the comb, is in fact often sold in this way, wrapped up in waxed paper or cellophane and if this can be done, the chief difficulty is overcome.

The trouble with sections is that, when packed in the super, they form a lot of little compartments, shutting off small companies of bees from the main body, and they dislike this so much that they will often fill the brood-chamber completely and swarm, rather than make use of them.

Having become familiar with the simpler side of the craft, the bee-keeper will, if he is at all keen, like to try and get sections; for the time may soon come, especially if we get a series of good honey years, when once again sections alone will command a ready market, secure from foreign competition. Comb honey is difficult stuff to transport and at no time has very much been sent to Britain from abroad, though there was a time when a fair amount was sent from California.

Sections, by which I now mean the wooden frames in which the comb is built, are made of fine white basswood, available in the USA. They reach the bee-keeper as flat strips grooved across in three places and dovetailed at the ends. There are several styles—four bee-way, two bee-way and no bee-way—

any of which may have plain tops and side grooves, ditto with no grooves, or split tops with or without grooves. Split tops are best, but there is not a pin to choose between the other forms. I have used them all and never discovered any special advantage in any one of them.

PREPARING THE SECTIONS

Each section is filled with a full sheet of specially thin foundation, which can be had either cut to exact size, or in strips from which three can be cut. Small strips of the same material can be used as 'starters' which reduces the outlay to a trifle, but it is not good economy, for there is no surety that the sections will be properly completed—they rarely are, if starters are used.

To fold sections, a frame or 'block' which exactly holds one is used, so as to ensure a perfect square. But first, to avoid breakage in folding, the cross grooves must be moistened. I find it best to take a batch, say ten, hold them firmly together, put them under the tap, and let water run down each groove. Stand the section in the block, grooves inwards, so that the split top projects and fix the half with the bevel on top sloping away from you. Fix the square of foundation on this bevel so that it is quite level and barely touches the bottom, sliding it down the grooves—if any. Press the top firmly against the bevel. This is important, for if it is not so secured, the sheet will probably buckle when the other half of the top is pressed down into its dovetails.

As completed, the sections are put in the racks and thin metal or wood partitions are placed between each row. The last row has a board pressed close against it by a spring. Too much care cannot be used to get sections square, with flat foundation and the dividers properly adjusted. Irregular bulgy sections are an abomination to handle afterwards.

From long experience I regard it as useless to put a section rack on early in the season. The best super to give first is always either a standard or shallow frame super and not until the main flow arrives and the bees are working eagerly in this, should a section rack be added. Bees can often be tempted into the section rack by putting one or two partly filled sections in the centre; unfinished sections from one season can thus be used for the next. If it fills rapidly, another can be added when it is half full. An empty section rack should never be put under a half-filled one, but when the bees have well started in the second, it is a very good idea to transpose them, for the first will then have less traffic over it and can be removed immediately it is seen to be well sealed.

In a really good season, this plan should result in a satisfactory crop, but in moderate seasons it will rarely procure anything like full racks. A plan I have found very successful in such conditions is to use a stock which has nearly filled two brood-chambers. When the main flow begins, I take off the top chamber and remove from both all combs except those containing brood and only enough of them to fill one chamber. The rest I give to weaker stocks in the apiary. All bees are shaken back into the brood-chamber.

Having put an excluder over the brood-chamber, I put on two racks of sections at once. Both will be filled with bees immediately and, given fair weather, they are filled rapidly.

Even so, you must be on guard. If a change for the worse in the weather takes place, the stock will be practically foodless, so it is a wise precaution to tuck away any combs filled only with honey, taken out at the start of the operation, and put them back on top at the first sign of a bad break.

If one does not care to risk trying for sections only, it is always possible to get a few in a shallow super, by using a hanging frame into which three sections can be fitted. At one time, racks holding sections in this way could be purchased, but I have not seen them lately. It is rather a fiddling method, but it is not difficult to fit three sections into an ordinary shallow frame, wedging up any gaps securely so that there is no room to build burr comb.

Sections should be removed as soon as they are completely sealed, using the super-clearer. I have, on occasion, when the honey-flow was in full blast, taken section racks off without it, pulled them to pieces and shaken the bees off, but I do not recommend it. After removal, clean and pack the sections into tins out of reach of any pests, including robber bees, without delay.

CUT COMB

A great many of the disadvantages of sections can be avoided, however, by the production of 'cut comb'. The method is to insert frames fitted with thin unwired foundation into supers between good, straight, fully drawn combs, preferably just before an expected nectar flow—if you can manage this. The bees will rapidly draw out, fill and cap the cells without displaying the reluctance so often found in the case of sections. When the combs are sealed, they can be withdrawn and cut into pieces about $2\frac{5}{8}$ in by $3\frac{1}{2}$ in. These will just fit into the plastic containers specially made for the purpose which are available from the appliance manufacturers. A well-filled piece of comb this size will weigh almost exactly 8 oz. Cutting tools specially made for the job can be had, but for the man producing only a small amount of cut comb for his own and his friends' delectation it is simple to do the cutting on a clean formica-covered kitchen table or a similar surface. Use a cold, sharp kitchen knife and very little waste or mess will result. In the case of heather honey, owing to the thixotropic nature of the honey, no seepage at all will occur.

An out-apiary sheltered in a disused chalk pit, showing two W.B.C. hives at the far left of the front row, four British National hives in the centre and, right, four Langstroth hives. The rear row contains eight Nationals

OUT-APIARIES

The bee-keeper who operates on a large scale must necessarily distribute his stocks, so as to have, at the outside, not more than 50 in one place. He will scarcely be able to make a living from less than 300 stocks, so he must have what are called 'out-apiaries'. Even the enthusiastic amateur may also find it desirable, if he lives in a town, to keep most of his bees in a more remunerative place than one surrounded by bricks and mortar.

Obviously there are advantages in this plan. Sites can be chosen which are in immediate proximity to large orchards, meadows growing abundance of clover or last, but not least, the heather moors. There are also formidable disadvantages. For one thing, it is essential to have one's own transport or, at all events, to be able to call on it at a moment's notice. A considerable risk will be run of losing swarms, unless it is possible to pay weekly visits during the season and forestall such disasters by examination. There is also the danger from marauders—human or animal. In fact, all the risks of a normal apiary are intensified if the bees are away from home.

SUITABLE SITES

It is not always easy to find sites for out-apiaries. Farmers, taking them on the whole, are not very fond of bees on their land. Still, given a friendly approach, it will generally be found possible—for a consideration—to rent a site which is not suitable for normal cultivation. Rough corners which it does not pay to plough, old gravel pits or quarries, a clearing in a wood or beside a copse or orchard, are often ideal places for bees. The fruit grower who recognizes the importance of bees in pollination and the growers of clover seed ought to be very willing to accommodate a few stocks.

The site should be reasonably accessible. If hives and supers of honey have to be transported over rough ground, or worse still, manhandled a considerable distance, it is easy to imagine what damage may be done and labour entailed. On the other hand, if the place is too easily reached, you never know who may come along and 'lift' the crop. Such happenings are not very frequent, but they do occur and it is well to be prepared. Whatever the nature of the site, it must be protected against intruders, particularly farm stock. Horses or cattle can soon upset hives and cause much trouble to all parties. The cost of a few stout stakes and barbed wire will be a fairly cheap insurance against such disasters.

When a townsman has all his bees in one out-apiary, it will be most convenient to have the extractor and indeed all apparatus on the spot, and in that case a honey house of some good pattern will be required. It will thus be possible to complete any operation at one visit. If the out-apiary is only one in a series, this will not be practicable. It would be an unwarrantable extravagance to have an extractor at each place, to say nothing of the difficulty of providing power. In some countries, notably Australia, large apiarists have an extractor mounted on a truck, deriving power from the engine, but it is much better to take the supers home to the main apiary. Even then, it is desirable to have a shed of some sort, if only as emergency shelter.

MOVING THE APIARIES

Apart from keeping bees in out-apiaries, there is the possibility of moving them to certain crops at favourable times. It may well be profitable to take the stocks to a field of clover or buckwheat in August, leaving them there until the blossom is over and bringing them back just as they are. This is not often likely to be worth the trouble, but taking bees to the larger fruit orchards in spring and to the moors in autumn is regularly done.

There is considerable difference between the two, for while the purpose of migration to fruit blossom is to fertilize the flowers, bees are taken to the heather in the hope of a crop of this most remunerative honey.

The prospect of sufficient crop to repay the cost of moving to fruit blossom is rather remote and it is recognized that the bee-keeper requires a fee for bees placed in orchards or on farm crops. They should, at least, be able to build up good populations and under exceptionally good conditions, a surplus may be stored, though few colonies have reached the strength necessary for heavy storage at this time.

SPRAY POISON

One grave danger exists: the risk of bees being poisoned by the spraying now regularly carried out in commercial orchards. There are numerous kinds of spray, not all detrimental to bees, but that put on just before blossom opens and again after it has fallen, is an insecticidal preparation, fatal to bees. Theoretically, trees should never be sprayed while bloom is open, but since all varieties do not bloom together it is difficult to avoid it entirely. Moreover there is danger that weeds, such as dandelions growing under the trees, may receive the spray. If poisoned pollen is carried home by bees, the brood is liable to be poisoned. If the nectar is affected, many bees will die before they can reach home.

To minimize the risk from this source, bees should not be taken to the orchards until after the bloom opens and must be removed before the post-blossom spraying. This means that the move will have to be made at a moment's notice, since spraying operations are chiefly determined by the weather. Even if these precautions are taken, risk is not entirely avoided, especially in a dry spring, for spray naturally falls on twigs and branches. A smart shower will wash it off, but in dry weather bees may suck dew from the boughs in the morning and absorb poison. This risk may be reduced by providing a safe clean water supply near the hives.

HEATHER TIME

Migration to the heather is commonly practised in Britain, for the honey season in other districts is over by the end of July, when heather is just coming into bloom and because clover is lime-loving and heather lime-hating, it rarely happens that both can be reached from the same site. Some large bee-men transport stocks hundreds of miles to take advantage of this late crop.

SHIFTING STOCK

Moving bees, as may be expected, is not a task to be undertaken lightly and, unless due regard is paid to certain rules, is liable to occasion trouble. As I have said, bees locate their home so exactly that they cannot find it if it is moved more than a few feet, and movement in the home apiary should be made by steps of not more than a yard a day. As such moves have often to be made, it is useful to have some convenient means of doing it. Hives are pretty heavy during most of the year and merely to lift one with supers on is as much as most people can manage single-handed; to carry one any distance is out of the question. To facilitate this moving, I use a simple sledge, made of four stout battens nailed across each other, so that the hive stands on the upper pair and slides on the lower, to which I affix a stout cord for hauling. Thus I only need to slide the hive on to the sledge at the first move and slide it off again at the end.

A fully loaded hive which is too heavy for one person can be carried by two people any reasonable distance, if two ropes are passed round it, one near each end and a couple of poles passed between cords and roof. The bearers can then lift the poles on to their shoulders.

In emergency, a stock can be moved to any new position at one go, by shutting the hive up for three days. In the evening, perforated zinc is put over the entrance and a board propped in front to shade it from sun. In hot weather ventilation must also be given by some other means, the best being to replace the usual cover by one of wire cloth. After three days' confinement, bees seem to lose their memory and they locate the new position, as a swarm will always do if put down on a new site in the same apiary. This is the best method to follow if the stock is to be moved anywhere within a mile. If the distance is more than a mile and less than two, it will suffice to stuff grass into the entrance before the move. The bees have to dig their way out and this seems to make them aware they are in a new spot, which they memorize before starting work again. Outside the two-mile radius, no precautions are necessary to ensure orientation of the new home. Needless to say, bees should never be closed in till all have ceased flying for the day.

SINGLE- *v.* DOUBLE-WALLED HIVES

It is in the matter of moving bees that single-walled hives score most heavily over the double-walled types, for not only do the latter make inordinate demands on space in a vehicle, but fixing the loose parts together is cumbrous and time wasting. Single-walled hives of the Langstroth type present no difficulty. Appliance makers supply clamps which fix the parts securely together, entrance closers of perforated zinc and a special ventilating cover. To provide ample ventilation is most important and the only satisfactory way is to replace the wooden inner cover by one of wire gauze or perforated zinc. This is fixed in a wooden frame which screws into the hive walls and, to prevent the combs from swaying, presses on their ends. It may be necessary to fix a strip of wood across the ends of the frames to ensure this. Its thickness will depend on whether the frames are flush with the top of the chamber, or have a bee-space there.

VENTILATION

It is as well to emphasize that bottom ventilation, however large, is not sufficient, for bees will crowd round it in their desire to escape and thus block it up. Full ventilation on top is the only safe plan, for while bees can combat cold by close clustering, their efforts to counteract heat only make matters worse. Their one desire is to get outside and on abnormally hot days they will often come out *en masse* and cluster round the sides till the heat abates.

In cool weather, the roofs can be put over the ventilating screen, providing it is raised by sticks or blocks to allow air to circulate freely. In hot weather, it is better to leave them off and provide some light covering to prevent the light from exciting the bees. Some bee-keepers have their hive floors without any projection in front and stand them in the upturned roof in the van. It is better to stand the hive so that the combs are parallel with the sides of the vehicle.

A light truck with sufficiently low platform is the ideal vehicle, but for most amateurs this will be a matter of adaptation. Some car owners have a trailer made specially. I have seen four or more hives packed in a car and its trunk, but it is very inconvenient. Whatever vehicle is used, jarring should be avoided by placing sacks of chaff or straw under the hives. Needless to say, the journey should be made at moderate speed and special care taken in cornering. Entrances to out-apiaries and moorland tracks need careful negotiation, especially at night. This is eminently the best time to move bees in summer and, since you must always wait till they have finished flying before closing them in, it is quite convenient for there is no loss of foraging time. Bees are soon at work next day on the new ground.

It is never desirable to send bees by rail in their hive. They should be transferred to a proper travelling box and the hive sent separately.

HEATHER HONEY IN BRITAIN

The production of heather honey requires a technique which differs considerably from that used in getting the crop from clover, and those who live in districts where the latter is of small account must manage their bees very differently. It is quite useless to build up a big population for the clover and hope that the same bees will gather as much from the heather, for not only will the clover foragers be worn out, but the peak of brood-rearing has so long passed that there are comparatively few young bees reaching the foraging stage.

In the days when skep bee-keeping was general, there was quite a large traffic in swarms, which were sent from the south to the north specially for the heather and it was generally considered that to unite a number of late casts was the most satisfactory way of providing stocks for the purpose. The development of bee stocks is, in any case, always later in the north and, as far as possible, breeding must be encouraged to continue considerably later than when clover is the sole object.

There are, of course, some areas like Dartmoor, the Surrey Hills, parts of the New Forest and a few other districts where heather grows in sufficient profusion, but here it can best be looked upon as a supplementary crop, valuable for providing winter stores, instead of being the main object as it is in real moorland counties.

Another handicap is that days are shorter and nights longer than during the clover flow. Sometimes, of course, August is a very hot month, but equally often it is wet and sometimes cold. For this reason it is seldom wise to rely on bees building comb at the heather, unless it happens to be one of the hot years, when they will build out foundation as readily as at any other time. It is far better to mobilize all built-out comb, whether standard, shallow or section, and take this along with the bees. It will always result in a larger surplus of heather honey.

Those who live in or near the moors will be in the best position to obtain sites: taking bees to the heather and bringing them home again will be part of their normal procedure. For those strange to the moorland country, living at a distance, the whole thing will be a gamble, for they will have to find suitable

An apiary temporarily moved onto the moors near Hadrian's Wall, England for the heather-flow

standing ground, either by a preliminary visit or by correspondence. Members of bee-keepers' associations will find it best to get in touch with the association secretary in the moorland district chosen. The cost of moving the bees out and home again will also have to be considered.

Again, those who live at a distance will most likely not be able to visit the hives often, if at all, unless, which is not at all a bad idea, they find quarters in the neighbourhood and make a heather holiday of the enterprise.

In these circumstances, all the supers the stocks are likely to need should be put on at the start. Excluders are not necessary, as there is very little danger of the queen needing more breeding space than she has in the brood-chamber. Two shallow frame supers or section racks will normally be ample and they can be put on before the move. In hot weather this will be an advantage, since it provides more ventilation. Needless to say there should be a fairly good supply of honey in the brood-chamber in case conditions should be bad, but combs overloaded with new honey would be dangerous on the journey in hot weather.

Once in a while, bees swarm at the heather but it is sufficiently unusual to be negligible and in due time the stocks can be brought back. If a good crop has been secured and the brood-chamber as well as supers filled up, the trouble of feeding will have been saved, but it is well to remember that there are some years when they are likely to return lighter than when they set out, in which case feeding will be necessary. It is therefore best not to delay too long in bringing them home.

EXTRACTION

Heather-honey press

Heather honey is quite different from any other. It is a thick jelly which cannot normally be extracted in the machine. There is a device which, by inserting a number of pins into the cells, temporarily breaks down the jelly so that it can be centrifugally extracted, but most people have found it a rather impracticable affair. It is far better to get the crop in sections, which are snowy white and always command top prices. Failing this, shallow frames fitted with thin unwired foundation can be used, and the comb cut into pound squares and wrapped in waxed paper or cellophane.

The recognized way of extracting heather honey is by a press. Several patterns are obtainable, but in each the comb, having been cut out of the frames and wrapped in muslin, is pressed until the honey is forced out and only a flat slab of wax remains.

13
Wild Bees

Even if he never noticed them before, the enthusiastic bee-keeper will soon discover that wild bees are fairly numerous. The term 'wild' is not strictly correct, for it is true that the honeybee is not a native of North America so that any bees found in unorthodox places have escaped from human control at some time. In good bee seasons, many swarms are lost, owing to the negligence or ignorance of their owners and, if not retrieved, find their way into more or less inaccessible places: hollow trees, chimneys, house walls and the like.

When a bee-keeper acquires some local reputation, he is sure to be called on from time to time to dislodge some of these uninvited guests and I know one or more 'experts' whose names are on the books of the local police for availability to deal with swarms which have settled in awkward places. Only last year one settled on a traffic signal and obscured the lights until a bee-keeper was found to remove it.

I need say little about the collection of swarms, for they can be dealt with by the same methods as are used in the home apiary, but it is a good idea, if you are likely to be called on to cope with such emergencies, to have a box ready with smoker and other things useful for the work, so that you can repair to the spot without delay.

SETTLED SWARMS

Swarms which have settled in a building and built combs are quite another matter. Each case must be dealt with on its merits, for they will never be exactly alike, but a few general hints based on experience may be helpful.

If the intrusion has been recent and no combs have been built, it may be possible to drive the bees out with smoke or carbolic. One expert states that he can get bees out of an otherwise inaccessible chimney by putting a comb of brood in the fire-place. The draught carries the smell of the brood up to the swarm, which descends and clusters round it.

A colony which has established itself cannot be shifted unless the combs can be reached and removed, and this often entails removing tiles, bricks, floors, plaster and what-have-you so that unless the bee-keeper is a specially handy man, he should not undertake the task without the aid of a competent workman. This means that the cost of getting the bees out is likely to be greater than their value. Now and again, it is true, a considerable amount of honey can be found in such a place. I remember one instance when every bath, bucket and basin in the house was filled with it, but generally speaking, there is not so much honey as rumour suggests.

DRIVING

'Driving' may sometimes be a useful accomplishment for those who wish to get bees out of such places. This process was evolved during the days of transition from skep to frame hive bee-keeping and, in its simplest form, is carried out as follows. The skep to be driven is first well smoked and then turned upside down. An empty skep is then fixed to one side of it, in the position occupied by a half-open lid, so that while the light shines full into the occupied skep, the empty one is dark. Having made the junction between the skeps secure by passing a skewer through both and fixing angle irons to hold the top one rigid, the operator thumps the sides of the full skep with the palm of the hands, choosing those sides where the ends of the comb are joined. Very soon the bees leave the combs and march up to the dark interior of the empty skep. When the bees are out, the combs can be cut out, those containing honey put into a basin and the brood set aside to be dealt with as will be described later.

There are few occasions when the whole of this technique can be used to get bees out of a building, but sometimes it may be possible to drive bees from a place where the combs cannot be uncovered into a clear space where they can be brushed into a box.

IN TREES AND WALLS

Hollow trees are quite commonly occupied by stray bees, but unless they are *very* hollow and sufficiently soft to permit good-sized holes to be cut to get at the nest, they can rarely be cleared out unless the tree is felled and the trunk split open. I once got a small colony from a trunk which had been lying in a lumber yard for a year. Nobody had known the bees were there when the tree was thrown and carted.

When the occupied place is a hollow wall, or under a floor, the first thing is to locate the exact spot. This is not always easy, for the entrance the bees are using may be some way from the nest. Sometimes the warmth can be felt through plaster or wood, and by placing the ear against the wall the spot can generally be detected by the hum.

When the place has been located, as much of the area as possible must be exposed by removing boards or cutting away plaster. By stuffing rags or paper into any empty space surrounding the combs, the bees should be prevented from wandering into more distant regions. The combs should be detached whole if possible and propped up in the collecting box complete with bees. If lucky, you may get the whole lot quickly and cleanly but, if the bees are truculent or excited, they may wander here and there and take up endless time. The key to the situation is, as usual, the queen, and if she can be found and put in the box with the comb, the box need only be left close to the spot and all the bees will collect in it by nightfall.

DESTROYING THE BEES

If, on inspection, it is found that the position poses too many problems, particularly high cost of repairing damage, the alternative is to kill the bees. If a quart or so of gasoline is poured right into the nest, it will quickly finish the colony, but every precaution against fire must be taken. A solution of cyanide of potassium may also be used, but this is not very effective unless it can either be injected right into the nest or concentrated in a confined area to keep the fumes in. In some cases this might be effected with wet sacks but, as I have said, one can only use common sense and treat the case as seems most suitable.

If the combs are not removed, it may not be long before a fresh swarm takes possession, unless every possible entrance is sealed. I suppose one could introduce a few wax moths and leave them to clear the place out, but it is hardly good bee-keeping to encourage this pest.

This reminds me that swarms can often be captured without trouble by leaving a hive standing with combs. The ethics of this have often been discussed in bee papers, some contending that it decoys one's neighbour's swarms, but it does not *cause* the swarms and if all bee-keepers kept such a bait, one would cancel out the other, and fewer swarms would go into inaccessible places. The real objection to leaving empty combs about is that they attract wax moths, so if this plan is followed—I leave each to judge of its merits—the combs should be put away immediately the swarming season is over and treated to destroy wax moth larvæ. Indeed, if old culled combs are used, they can and should be melted down immediately their purpose has been served.

Taking it by and large, bee-hunting of this kind is more sport than bee-keeping. Mr Bee-Mason, who has since obtained an international reputation in other spheres, used to revel in it and some of the first and finest bee-keeping films were made during some of his adventures.

DEALING WITH A COLONY

Supposing you get a colony from one of these places, what should you do with it? If there is the smallest doubt about its health, it should be destroyed at once, for it might prove a very costly addition to the apiary. If there is no doubt of its vigour, it can be hived on foundation or combs like a swarm and treated according to requirements of the season. If there are good slabs of brood, they can be trimmed to fit inside a frame and tied in with tape, which the bees will, in due time, cut off, since by then the connections with the frame are secure. A frame of foundation should be put in the centre for the queen to lay in and as the brood empties from the old combs, other frames can be inserted and the old comb removed.

Even if the kind of bee is inferior to your own, such colonies can be used to build up weak stocks or requeened from the best stock, used as nuclei and so on, and this is just as likely to produce a good stock. I once got one with extremely little trouble and took 80 lb from it the next season.

TRANSFERRING FROM TEMPORARY QUARTERS

It sometimes happens that, from want of proper hives, bees are put into boxes or skeps as a temporary expedient and either forgotten or neglected, so that they establish themselves fully. To get these into a proper hive is not difficult. Having succeeded in wintering the stock safely, transfer it in the spring to a frame hive in this way. Put six frames of foundation in the middle of a brood-chamber and on each side place a division board. Get a piece of tough material, say American cloth, and cut a hole in the middle about 6 in across. Lay this over the frames and stand the box of bees over it. As the spring advances, the bees will work down through the cloth until the queen is laying in the frames below. As soon as this is the case, an excluder is put on and the box returned and allowed to remain until it is filled with honey, when it can be removed. If necessary, extra frames are added below.

14
Anatomy and Physiology of Bees

Although a knowledge of anatomy is not essential to a bee-keeper, some acquaintance with the mechanics of a bee's body is at least as desirable as it is for a motorist to understand the construction of the various parts of a car.

Insects belong to the invertebrate part of the animal kingdom. Instead of being built up on a bony framework, their organs are attached inside a case made up of twenty rings or segments. These segments are distributed in three groups—six in the head, three in the thorax and eleven in the abdomen. In the course of evolution, some segments have become fused together, so that they are not very distinct and it is usually only in the larval stages that they can be readily distinguished. Even in these, only thirteen segments are readily apparent.

The substance of this external framework is not bone, but a peculiar material called *chitin*, which is tough, flexible and more or less reinforced by a much harder substance, *sclerotin*. In the larval skin, sclerotin is absent, as it is also in the flexible membranes which join the segments together in the adult. A black pigment, *melanin*, is also found more or less in the adult, giving, according to its amount, the range of colouring familiar in the full-grown bee.

All outgrowths from the main structure are formed of the same material; legs, wings, antennæ, spines or hairs have varying quantities of these substances, according to the purpose they serve. Some of these outgrowths are rigid, others are articulated to the body by a flexible membrane, which permits them to be moved freely. Certain parts of the head and thorax are strengthened by braces and buttresses made of the same material as the exterior wall.

The body of the bee is divided into three well-defined parts, separated from each other by narrow connections. These parts are the head, containing the chief sensory organs; the thorax, which supports the organs of locomotion; and the abdomen, in which are the digestive and reproductive organs.

HEAD

The head, in which six segments are so completely fused together as to be indistinguishable, is, when seen from the front, an equilateral triangle standing on its apex. Its most conspicuous features are the compound eyes which lie on either side. These are very much larger in the drone than in queen or worker, making its frontal aspect almost circular. In the centre of the head, between the eyes, are the antennæ or feelers, slender jointed structures consisting of twelve segments in the worker and queen and thirteen in the drone.

The first segment (scape), which is about a quarter of the full length, is articulated to the head in a socket, so that it can move freely in any direction. The remaining joints, which are approximately of equal length, are collectively called the flagellum. The first joint is articulated to the scape, but there is only slight flexibility between the other joints. The antennæ are freely covered with sensory hairs and, although precise proof of their functions is wanting and the belief that they are capable of transmitting ideas from bee to bee as humanity can by speech is probably an exaggeration, they are manifestly sensitive tactile organs, and also the chief seat of the sense of smell. At least seven different kinds of sensory organs have been defined on the antennæ and they are so numerous that nearly half a million are to be found on one antenna of the drone.

On top of the head—rather more forward in the drone—are three *ocelli* or simple eyes arranged in a triangle. Unlike the compound eyes, which will be discussed later, these ocelli have very limited vision and probably can do no more than distinguish light from darkness.

MOUTH

The mouth is one of the most highly specialized and elaborate in the insect world, in which this important organ varies to a remarkable degree, according to the nature of the food. It comprises the *labrum* or upper lip, which bears on its inner surface a delicate membrane called the *epipharynx*, considered to be an organ of taste. On either side of the labrum, attached by ball and socket joints, are the *mandibles* or jaws, which move from side to side. They are hard and powerful and in the queen and drone are notched, but those of the worker are smooth on the inner side. Their principal use seems to be the manipulation of the wax and propolis used in building, but they are also employed to remove loose portions in the hive and can chew wood and other fibres. Each mandible is served by a gland, which provides a fluid that mixes with the wax while it is chewed. The proboscis is formed of a number of parts, which correspond to the jaws or *maxillæ* of biting insects, but have been remarkably adapted to serve the purpose of a sucking tongue. Its central part is the *glossa* or *ligula*, a narrow strap six millimetres long in the worker, but much shorter in the queen and drone. It is covered with sensitive hairs, has a central groove on the underside and a round concavo-convex portion at the tip called the *labellum*, sometimes known as the spoon. Subtending the glossa is a pair of short lobes, the *paraglossæ*. These are flanked by the long four-jointed *labial palpi* and external to these are the greatly modified maxillæ, the chief part being the flat leaf-like *galæ*, which extend almost as far as the labial palpi.

All these parts are very mobile, being actuated by powerful muscles. When the bee feeds, the glossa is projected to and fro, collecting the liquid on its hairs, whence it is swept up towards the mouth. The labial palpi, maxillæ and epipharynx co-operate by closing round the glossa, to form a tube through which the liquid is drawn into the gullet.

THORAX

Behind the neck which separates it from the head is the thorax, consisting of three segments fairly easily recognized and named respectively, *pro-*, *meso-* and

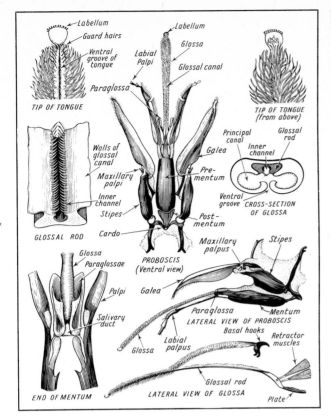

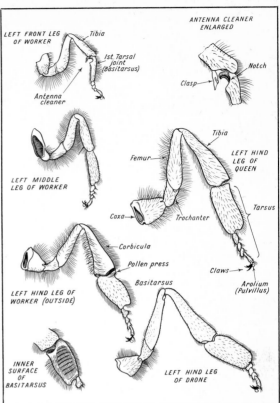

meta-thorax. The pro-thorax contains the first pair of legs, the meso-thorax the second pair and the first pair of wings and the meta-thorax the hind legs and hind wings. To this portion is also closely fused the *propodeum*, which in other insects forms the first of the abdominal rings.

Above left: details of tongue
Above right: legs of the honeybee

LEGS

Each of the six legs is divided into five major portions. At the base, the *coxa* or hip, next the short *trochanter*, jointed to the coxa, but fixed to the much longer *femur* or thigh, which is jointed to the *tibia*. Finally the *tarsus* or foot. This has five joints, the first being as long as the other four combined. At the end of the tarsus is a pair of hooked claws and between them a fleshy pad (*pulvillus*) which secretes a viscous fluid, enabling the bee to walk on a surface to which its claws cannot be attached. The claws play an important part in the life of the colony, for it is by hooking one bee's legs to another's that a swarm is able to suspend itself in the familiar mass. All the legs are more or less covered with hairs and spines, some of which are highly specialized for important functions.

On the forelegs, which are the shortest, there is a semi-circular opening (*sinus*) at the upper end of the *palma* or first tarsal joint. This sinus is inwardly fringed with stiff hairs, resembling and indeed serving as a comb. Corresponding with this is a spine on the end of the tibia, bearing on its inner edge an angular projection (*velum*), which fits into the sinus at the will of the bee. The purpose of this comb (*pecten*) is to clean the antennæ, which are liable to be fouled with pollen, honey or other substances. The antenna is laid in the sinus, which is then closed with the velum and the antenna is drawn through repeatedly until effectively cleaned. The second pair of legs carry no special organ, but the tibia bears at its end a simple spine such as is found in this place in many insects. This removes wax scales from the wax glands.

The hind legs of the worker are the largest and are provided with special apparatus for the collection and transport of pollen, known as the *corbicula*. The outer side of the tibia is hollowed out and its edges fringed with stiff recurved hairs, which retain pollen packed into the hollow. The first tarsal joint (*planta*) is hinged to the tibia at its inner edge, so that it will open and close freely and the top of the outer edge is broadened and slightly hollowed, so that it acts as a movable bottom to the corbicula and by pressure from below, the pollen mass is steadily pushed upwards until the corbicula is full. The planta is provided on its inner side with transverse rows of stiff spines; with these pollen is combed from other parts of the body and transferred to the corbicula. All the legs co-operate in this work, pollen on the right foreleg being transferred to the left middle leg and from the left middle leg to the right hind leg and *vice versa*. Neither queen nor drone possesses this pollen-collecting apparatus.

WINGS AND ABDOMEN

The wings are constructed of extremely thin membrane in two layers, above and below a framework of tubes which gives them rigidity and at the same time carries a supply of blood and air. The outstanding feature, shared by other members of the order, is that the two pairs can be united or disconnected at the will of the insect. On the front margin of the hind wings—which are scarcely more than half the length of the fore wings—there is a series of minute hooks, which correspond with a fold on the hind margin of the fore wings. When preparing for flight, the bee fastens these hooks together, so that one broad plane is formed and the two pairs act as one. On alighting the bee detaches the hooks and folds the wings back against the body. The value of this to a creature which has to go in and out of very narrow cells is easily appreciated.

In the adult bee the abdomen is reduced to six visible segments—seven in the drone: these overlap slightly and are connected by thin membrane which allows considerable expansion. Each segment is composed of two plates, the upper or dorsal plates larger and curved under, so as to overlap the lower or ventral plates, which are slightly raised in the centre. On either side of this ridge is a transparent pentagonal area, through which the wax secreted by glands beneath them exudes and forms a solid scale on the surface.

The sting and genitalia carried at the end of the abdomen will be described later (see pages 142 and 144).

NUTRITION

Respiration and circulation are carried on in insects in almost the opposite manner to that of vertebrate animals, for instead of the blood being carried to the air, air is carried to the blood, which is a clear, yellowish fluid, containing various forms of corpuscles or hæmocytes. It does not carry oxygen to the organs, as in vertebrate animals, since that function is performed by the tracheal system; instead it distributes the digested food and carries off waste products and carbon dioxide.

There is no system of blood circulation at all comparable with vertebrate arteries and veins, but a central tube runs the whole length of the body above the alimentary canal. This tube is closed at the abdominal end, but open at the head. Divided into five ventricles, each communicates with the other by a valve which opens in a forward direction. Blood is received through inward opening valves in the side of the ventricles and is continually propelled forward to the opening in the head, whence it flows round into every part of the body.

RESPIRATORY SYSTEM

There is no central respiratory organ comparable with the lungs of mammals or gills of fishes. Instead of this, an extensive system of tubes (*trachea*) traverses the whole body. In structure they resemble a coiled spring, but this appearance is due to folds of the chitinous membrane. Trachea are, in fact, an internal extension of the outer integument and during the moulting which takes place at intervals in the growing period of an insect's life, they are cast off with the outer skin.

The trachea commence at openings in the side of the body called *spiracles* of which there are three pairs in the thorax and seven pairs in the abdomen of queen and worker. The drone has an extra abdominal pair. A fringe of hairs before the opening filters the air as it enters and there are also lips with which the aperture can be entirely closed. Widest at their commencement, the trachea diminish in size to infinity in the course of their ramifications, but in certain places they are expanded into thin-walled air-sacs. Very large ones are found in the fore part of the abdomen and the meta-thorax. Such sacs are found in all flying insects, but are particularly well developed in the bee. When filled with air they diminish the specific gravity and increase muscular activity by increasing the air supply.

DIGESTIVE ORGANS

In sucking insects, the alimentary canal begins at the tip of the tongue, through which alone food can enter. It continues as a narrow tube (*œsophagus*) through the head and thorax to the fore part of the abdomen, where it is enlarged into a crop, called in the bee the *honey-sac*. It is succeeded by a much wider mid-intestine or *chyle stomach*, where the chief work of digestion is carried out, and nutrient matter is absorbed into the blood through its walls. Between the honey-sac and the chyle stomach is a muscular organ, known as the *proventriculus*. Its function is merely valvular, so that it enables the insect to retain the food in its honey-sac or pass it on for its own use.

From the mid-intestine, any undigested matter passes into the hind intestine through an opening controlled by sphincter muscles. At this point there appear the openings of numerous small tubes, the Malpighian tubules; the free ends lie in the body chamber, from which they withdraw waste matter formed in the blood to discharge it into the hind intestine, which continues to absorb nourishment from the food until it reaches an enlarged portion (*colon*) where waste matter accumulates to be discharged at convenience. During winter, the fæces are retained in the colon, no matter how long the period of confinement, until an opportunity for flight occurs, when the colon is emptied through the *anus* during flight.

Salivary glands, situated in the head and pro-thorax, assist in the work of digestion and other digestive secretions take place in the stomach. The former belief that brood food was given from the chyle stomach has been abandoned and it is now clear that the food of both queen and larvæ is a secretion from the pharyngeal glands in the head, supplemented in the case of older worker and drone larvæ by honey and pollen from the honey-sac. Some investigators think that brood food varies in constitution according to the age of the larvæ and that the royal jelly given to queens differs from that supplied to worker larvæ.

SENSATION

Throughout the animal world sensation is controlled and directed in much the same manner by a multitude of lines or fibres called nerves, which lead to and from centres composed of receptive cells. In the simpler segmented animals, each segment has its own nerve centre, consisting of a pair of *ganglia* connected to each other and those before and behind by nerve fibres. As animals become more specialized, there is a tendency for nerve centres to be reduced in number and concentrated in special regions.

In the bee, this chain of ganglia runs up beneath the alimentary canal. As may be expected, the largest is found in the head. This is the *brain* and unlike the others it is above the œsophagus. A smaller ganglion lies beneath the œsophagus. A pair in the pro-thorax serves the fore legs and two united pairs of rather larger size in the meta-thorax deal respectively with the middle legs and fore wings and the hind legs and hind wings. Five smaller ganglia in the abdomen supply nerves to the various organs there; the last two, rather larger than the others, serve the genitalia and sting.

The effect of this distributed nervous system is that each part of the body functions more or less independently. If we cut off the head, the body can run about, move its wings and continue its vital processes for a long time. If the abdomen is severed, the insect will continue to suck up liquid quite oblivious of the fact that as fast as it is taken in, it runs out behind.

Investigation into the special senses of the bee is very difficult and in regard to many subjects no definite conclusions have been reached. It would seem certain that the dominant sense is that of touch. This, in the peculiar propinquity with which the workers carry out their labours, must be highly developed. Sensitive tactile hairs are found in all parts of the body, while special organs like the mouth or the sting have feelers or *palpi* designed to guide their actions. Above all, the antennæ, so constantly in use and so meticulously cared for, are certainly capable of examining objects of all kinds with an accuracy of which we can scarcely form an idea.

SMELL

The antennæ, again, are certainly the seat of the olfactory sense, which probably comes next in importance to touch in the sense life of the bee.

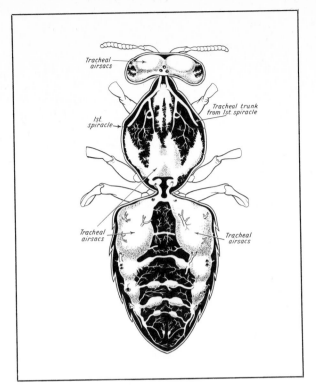

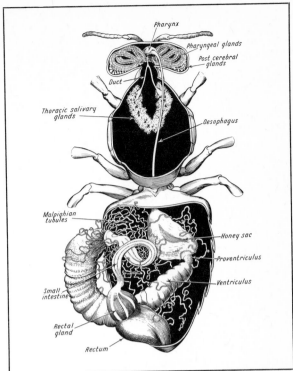

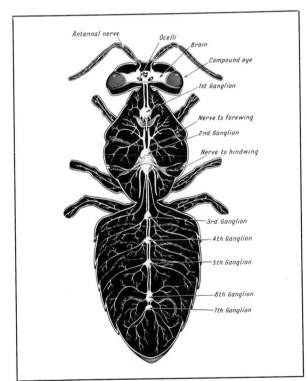

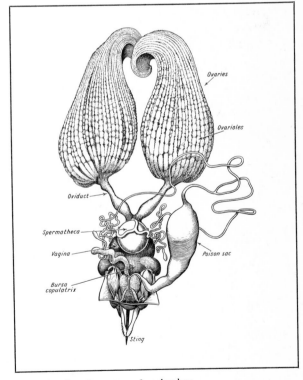

Top left: respiratory system of worker bee   *Top right:* digestive system of worker bee
Bottom left: nervous system   *Bottom right:* reproductive organs and sting of queen

Amputation of the flagellate part of an antenna destroys olfactory perception in proportion to the amount removed. Apart from the necessity of a well-developed sense of smell in insects which seek their food in flowers, this plays a great part in the domestic life of the bee. The presence of a queen on the wing is known to the drones by this, and the olfactory receptors on the antenna of a drone have been computed at 30,000, whereas the worker has only 5,000 and the queen 3,000. The recognition by individual bees of their own colony is also attributed to smell and this is especially associated with an organ in the abdomen known as Nassanov's gland. It is noticeable at swarming time that bees alighting on the spot where the queen is located elevate the abdomen so that this organ is exposed, intensifying and distributing its odour to the flying bees.

The sense of smell and taste are so intimately connected that it is probable that the same organs function for both purposes, but the tongue, the antennæ, palpi and tarsi are, in various insects, known to serve the purpose of gustation. In the bee some sense of taste is ascribed to the epipharynx on the labrum.

SOUND RECEPTION

The most obscure subject of insect sense is that of sound reception. It is clear from many experiments with hairy caterpillars that some at least of their hairs are sound receivers and since there are in many insects special organs for making sounds, mostly in connection with mating, it seems certain that sounds can be received and distinguished. In the hive bee, no definite location of this kind has been determined, though some have credited the antennæ with this sense in addition to those of touch and scent.

SIGHT

No such difficulty attends the sense of sight, for the eyes are obvious and their structure has been accurately investigated. I have already mentioned the simple eyes found in the centre of the head. The compound eyes are even more conspicuous, as they occupy, particularly in the drone, a very large part of its surface. Each eye consists of a great number of hexagonal facets, each of which functions as a separate lens. The number of facets varies from 3,000 to 6,000 in the queen and worker to as many as 13,000 in the drone. Arranged on both sides of the head, each facet reflects a slightly different picture from the others, so that the combined facets cover a large area though each contributes only a small fraction of it. The image thus formed on the brain is a mosaic built up from as many points of light as there are facets.

It is probable, from the very nature of their activities, that the eyes of bees are superior to those of most insects. It seems impossible to imagine how they could otherwise locate their hives with such accuracy, or pass from flower to flower so speedily. That they have definite colour perception has been clearly shown, but their keenest vision is at the ultra-violet end of the spectrum, where their range exceeds that of man. To the red end they appear to be blind.

PHEROMONES

Pheromones are minute traces of chemical substances that carry information so that insects can communicate by using a code of chemical stimuli. In a sense, they are similar to hormones but, whereas hormones are secreted internally and produce their effects in the organism in which they originate, pheromones are glandular secretions emitted externally which convey information to other individuals. They elicit reactions which modify the behaviour of the receptor.

In the honeybee two sexual pheromones are produced in the mandibular glands of the queen. She is constantly groomed and licked by workers and in the course of normal food sharing this so-called 'queen substance' is rapidly spread throughout the whole colony. A constant supply is necessary to maintain the emotional equilibrium of the colony and this also inhibits the building of queen cells.

When a queen ages, or if she begins to fail for any other reason, her supply of queen substance dwindles and a feeling of restlessness pervades the colony: the inhibitory effect is thus lost, with the result that either the queen is superseded or swarming takes place. In either case, the first result is the urge to raise queens. Queen substance also acts as a sexual attractant and as an aphrodisiac. It attracts drones to a virgin queen and incites them to copulate. Using a synthesized substance, research workers have succeeded in obtaining the copulation of drones with model queens fashioned in wood and coated with queen substance.

Other secretions act as alarm pheromones which provoke aggressive behaviour when they react with insects of the same species.

A third group are 'trail-marking' pheromones which are deposited by bees when returning to the nest after having discovered a source of food. The marked trails are then used by foragers recruited by dancing. Much research remains to be done on the whole question of chemical stimuli with obvious possibilities in the field of pest control.

DANCING

It seems that the majority of foragers do not search for food at random but, once a good source has been discovered, the information is passed to large numbers who exploit it. This had been suspected as long ago as 1788, when Father Spitzner suggested that the information was in fact passed to other foragers, but it was left to Karl von Frisch to develop his theory of the dancing bees over one hundred years later.

According to von Frisch, the successful forager returning from a source within 100 yards of the hive performs a simple 'round' dance on the face of the comb. She runs in small circles in one spot on the face of a vertical comb alternately to left and right. While she dances, a small group of workers gathers round her excitedly and touches her with their antennæ. She will regurgitate small drops of nectar from time to time which they collect in order to get the aroma and the taste of the new supply.

With more distant sources, the round dance changes to a 'wagtail' dance. This is something like a figure of eight, or perhaps two semi-circles made in opposite directions with a straight run between. The straight run is the important part. During this, the forager wags her abdomen vigorously. Distance is indicated by tempo: the slower the dance, the farther the distance. Allowance is made for head or tail winds so that an indication of energy required is conveyed rather than the distance in yards. Direction is shown by deviation from the vertical: for example, a 'wag' run 40° to the left of vertical indicates a source 40° to the left of the sun. A very great degree of accuracy is obtained by this method coupled with the tracer pheromones.

REPRODUCTION

The usual procedure is for drones to be raised in spring and, by the time these are mature, females capable of mating have been produced. When the first female is ready to leave her cell, the division of the colony takes place, and the old queen leaves the hive with a body of workers to settle in a new home. Shortly after, the first young queen leaves her cell and, if she is not prevented by the workers, proceeds to destroy all other queens in the cells. If the hive is still thickly peopled, the workers may guard the unhatched queens and the one at large will then leave the hive with a number of workers and found another colony.

The process is subject to considerable variation, owing to the characteristics of special races, weather conditions, etc. When a swarm has been delayed by bad weather, queens may become fully mature in their cells and able to fly immediately they are released, when they may accompany the prime swarm, or lead out casts, but sooner or later, one young queen is the only perfect female in the hive and she reaches the stage when she desires to mate. Having reconnoitred the neighbourhood so that she may return safely to the hive, she flies off, meets the drone and union is accomplished in the air.

THE MALE

The male sex organs consist primarily of two *testes*, which attain their greatest size during the pupal stage. They are composed of numerous *follicles* in which germ-cells or *spermatozoa* are developed. As the insect becomes more mature, the sperms leave the testes and pass through the ducts (*vasa deferentia*) into *seminal vesicles*, leaving the testes shrunken and empty. The vesicles merge into a single tube, the *ductus ejaculatorius*, into which two mucous glands send a secretion which mingles with the spermatozoa and forms a mass, *spermatophore*; this occupies an expansion of the duct known as the *bulb*. During the act of mating the spermatophore is discharged into the *vagina* of the queen.

THE QUEEN

The organs of the queen consist of two ovaries which, in a well-developed fertile queen, occupy a very large part of the abdomen. They are composed of nearly 200 tubes or *follicles* in which the eggs are developed. These tubes all converge to an *oviduct*, the pair of oviducts again combine to form one common passage into the vagina, which bears a pair of swellings, *bursa copulatrix*, that receive the horns of the drone. Near the point where the oviducts unite there is a round vessel, the *spermatheca*. This has its entrance into the oviduct controlled by powerful muscles. During coition the spermatophore ejected by the drone is forced into the oviduct which enlarges to receive it. The sperms then escape from the spermatophore and find their way into the spermatheca, where they are retained. Many thousands in number, they normally serve for the lifetime of the queen. By action of the controlling muscles, when eggs are passing through the oviduct, sperms may be released or not. Those which pass out without receiving sperm become males. When the spermatheca is allowed to release some of its contents, one or more sperms enter the egg through the minute opening at the top called the *micropyle* and a female bee is produced. F. J. Manning (*The Microscope*, Vol. VII, No. 11) considers it doubtful that the micropyle is the entrance for the sperm. He has observed as many as twenty sperms clustered at the end of the egg and regards it as likely that more than one passes through the egg membrane before it toughens. Only one sperm reaches the pronucleus of the egg, while the remainder perish in the plasm.

THE EGG

As laid by the queen, the egg is a white, slightly curved cylinder about $\frac{1}{16}$ in long, with rounded ends, the top slightly larger. The experior shell, *chorion*, is sculptured with a reticulated pattern and is lined with a thin membrane. The egg is stuck to the middle of the cell bottom in an upright position, but during the three days which elapse before hatching, it gradually falls over until it lies on the cell bottom.

On the fourth day, the larva hatches and lies curved and transparent-looking, surrounded by the milk-like food given by the nurse bees. It increases daily in size, curving more and more until, on the eighth day, it completely fills the cell diameter and is so closely coiled that head and tail meet. It is then shining white and destitute of limbs, but the thirteen segments including the head and the spiracles on one side are clearly visible.

Each day the supply of food is increased, but on the fourth day after hatching honey and partly digested pollen are substituted for the glandular food. This applies to worker and drone larvæ. Queen larvæ on the other hand receive glandular food lavishly throughout the feeding period.

On the ninth day from deposition of the egg, the larva ceases to feed and is sealed in its cell by the workers. Inside, it begins to change its position until it finally lies extended in the cell with its head towards the mouth and spins a *cocoon* of light silk. During the next twelve days the changes take place which transform it from the segmented legless larva into the adult, winged bee: the divisions between head, thorax and abdomen appear and the various organs become apparent. At first the white larval colour persists, but pigment is gradually formed and causes the skin to darken, though it does not attain full coloration until some time after the insect has left its cell. On or about the twenty-first day, the fully developed worker cuts a circular opening in the cap of the cell and crawls forth, ready soon to take part in the indoor work of the hive. During its rapid growth, the larval skin, which is incapable of expansion, is cast off five times and the pupal envelope is also discarded when the transformation is finished.

The time taken for this development is rather longer in the drone, where feeding is continued to the tenth day and the pupal stage lasts between two and three days longer. On the other hand, a queen attains full development in about fifteen days, and the pupal period lasts little more than half the time taken by the worker.

THE NEW WORKER

The newly emerged worker spends her first few hours wandering about the comb. She then begins to feed freely from the open honey cells. This appears to stimulate the glands which produce larval food and she begins to supply this to the brood. This stage is followed by one of plentiful wax production and comb building and it is not until about a week or ten days from emergence that the young bee takes its first flight. During the season numbers of these young bees come out together about the middle of the day and circle the hive in company, in order to locate its position and thus prepare for the strenuous outdoor life they lead in the summer time.

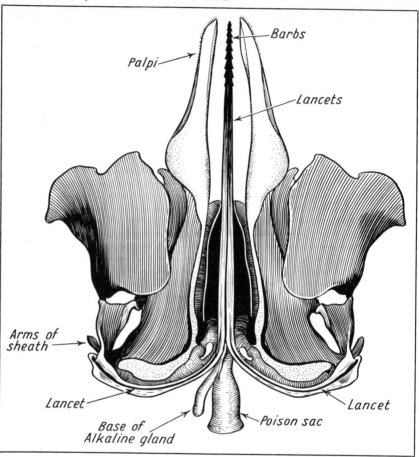

The sting of the worker bee

DEFENCE

The bee's sting consists of two fine tough *lancets*, lying side by side with a central channel, along which the poison is driven out through openings between the recurved teeth or *barbs* at the end of the point. These lancets pass between a hard sheath, which holds them rigidly in place, to arms that curve outwards to a point where they are attached to levers which cause them to revolve and drive the lancets down with enough force to penetrate hard substances. The movement automatically draws poison from the sac situated above the sheath. The poison is derived from paired glands, one of which secretes an acid and the other an alkali. Neither of these substances alone is toxic, but the combination is a powerful irritant, and causes severe inflammation and swelling. Palpi subtend the lancets and serve to test the attacked surface and ascertain its vulnerability.

Owing to the presence of the barbs, the bee is generally unable to withdraw the weapon after use, though if left undisturbed, it sometimes can, by turning round and round, manage to do so. Usually the whole sting with the poison sac is torn away from its body and by reflex action continues to pump poison into the wound. Hence the advisability of removing it as soon as possible, by scraping it off so as not to squeeze the poison sac.

The queen's sting is somewhat longer than those of the workers, slightly curved and the barbs are fewer and finer, so that she is generally able to withdraw it after use. Drones do not possess a sting, which is a purely female organ and is, in fact, a modified ovipositor. It is highly developed as such in some other insects, notably the Ichneumons flies, who use it to inject their eggs into the bodies of the larvæ on which they are parasitic.

Indeed modifications of this apparatus are almost endless in this order of insects. The wood wasp, for instance, can pierce hard wood in order to bury her eggs as much as an inch below the surface. The sawflies have, as their name suggests, a toothed weapon for cutting into leaf tissues, and the gallflies not only pierce the plant tissues, but inject a fluid which starts structural changes in them, causing abnormal and sometimes fantastic swellings.

Most interesting, because it indicates how the two functions of stinging and egg-laying have developed, is the case of solitary wasps which collect caterpillars or spiders and pack them into cells for the use of their young; they not only deposit eggs with the victims, but inject poison which keeps them immobile, but still alive, until the larvæ hatch.

15
Loss of the Colony, Pests and Diseases

It would be difficult to estimate the normal duration of a bee colony. Although the life of workers is measured in weeks, their loss is more than made good by rapid breeding. Queens live for several years, but one seldom dies of old age because the workers are prone to supersede her by a new one if she shows signs of failing. If the combs become inferior by constant use for breeding, the bees may cut them down and renew them. So, by constant renewal of worn-out parts, the colony might continue indefinitely.

Right: brood comb from a colony that died of starvation. The darker cells were once occupied by worker brood
Far right: a mouse nest between combs

STARVATION

In practice, many things tend to shorten the life of stocks and hundreds pass out annually for various reasons. Most of them die during winter and two causes account for by far the larger number. The first is starvation. Colonies which have only secured a modest surplus during summer, or have had most of their gains removed, cannot get through the long nectarless period between October and April unless artificially and generously fed: 10 lb of food is the absolute minimum required during winter, and this is not nearly enough to provide for breeding plenty of young bees to take the place of the old ones which die quickly during hard spring labour. Long experience has enabled bee-keepers to fix 50 lb as the amount needed to carry a stock safely through winter and early spring.

The indications of death from starvation are: no food left in the cells, many of which contain bees that have died in a last attempt to get food; bees dead on the floor have a shrunken and empty abdomen and the tongue is often extended.

LACK OF A QUEEN

Queenlessness is the second most frequent cause of death. If this occurs in summer, when there is plenty of brood, bees requeen themselves, but if a queen is lost at the end of the season, although a young queen may be raised, she may not be able to mate because of bad weather or absence of drones. In this case she becomes a drone breeder, and although the stock maintains some strength through winter, it soon dies out in spring as the old workers perish.

Signs of death from queenlessness are absence of brood or only drone brood in worker cells.

BAD WEATHER

Adverse weather conditions account for a smaller amount of loss. Strong colonies in a sound, undisturbed hive are seldom affected by the coldest weather, but a defective hive, which permits a strong draught to pass through, may be fatal in a severe winter. Cold is not so dangerous as damp, and a frequent cause of death is a leaky roof, which keeps the interior constantly wet. Before winter begins it is important to see that roofs are completely water-proof. When colonies die from this cause, the bees are heaped on the floor and often covered with mould.

INADEQUATE VENTILATION

The same result may be produced by inadequate ventilation. There must be sufficient circulation of air to carry off the moisture produced by the bees themselves. A wide entrance should always be provided at the bottom and the top coverings must either be porous, so that excess moisture can pass off constantly or, if a wooden impervious cover is used, the central feed-hole must be left open, or covered only with perforated zinc; above the cover, adequate space must be left under the roof to ensure movement of the air. Many bee-keepers find it better to close the lower entrance and provide one at the top during winter, so that moisture-laden air can pass off, as it will naturally do, owing to the interior heating produced by the bees.

MICE

Mice are occasionally the cause of winter loss. If the entrance is large enough for them to creep through, they eat the comb and agitate the bees, and this causes them to consume food at unsuitable times, with resulting dysentery and death. There is no excuse for such an event, for both good ventilation and mouse exclusion can be secured by having the full entrance open and a piece of excluder zinc tacked along the front.

POISONING

Most winter losses may be regarded as accidental and can largely be avoided if reasonable care is taken. Losses at other times are more likely to be due to disease, but in some years poisoning is prevalent. This is usually attributable to spraying of farm crops to kill caterpillars. It is said that if this is done at the right time—that is before the buds have opened and the crop has come to fruition—bees are not injured, but recent experience has proved that this is not necessarily true. When spraying is done, not only do buds and flowers receive the poison, but it is distributed over the stem and leaves and also on weeds growing under the crop. The bees may gather poisoned pollen from these weeds or, if a dry period follows the spraying and a humid night causes a deposit of dew on leaves and stems, bees will be apt to collect it freely.

Poisoning is indicated by the rapid disappearance of adult bees. Some never return to the hive and others die as they reach the entrance, but there may be numbers of bees on and around the alighting board with swollen bodies and every sign of paralysis. Almost fully grown larvæ may also be found outside, for if poisoned pollen is brought home, nurse bees and larvæ may be killed.

The only way to avoid this danger is to move the bees right away from the sprayed area, but the risk of poisoned moisture being gathered may be reduced by providing supplies of clean water near the hives. Attempts have been made to find effective sprays which will either be innocuous to the bees or so objectionable to them that they will not collect nectar or pollen containing them but so far no positive solutions have been found.

PRECAUTIONS AGAINST DISEASE

Bee diseases are divided into two groups: those which affect adult bees and those attacking the brood. Many bacilli and other germs are found in the bodies of bees under various circumstances. Some are definitely associated with specific complaints; others, though present in certain diseases, are not proved to have any real relation to them. Many micro-organisms are found in both healthy and diseased creatures.

Even when a disease has been clearly diagnosed, treatment is not always easy. The principal difficulty is to achieve isolation of the sufferers. To remove them from the hive for treatment is impossible, since bees cannot live long away from the colony. Nor is the isolation of an affected hive very practicable. It is possible to keep it closed for a few days, but the beneficial effect of a drug administered is then offset by unhealthy conditions, including cessation of breeding due to the enforced confinement. Attempts have been made to surround the affected stock with fine netting, so that only a limited radius of flight is available, but large numbers of bees perish from their unavailing efforts to escape from prison.

The consequence is that, although success may attend the attempt to cure a colony, since they are free to wander where they will, the bees convey infection to stocks in the vicinity. On this account it is often suggested that the only treatment for bee disease is the destruction of the affected colony. This is not a very agreeable solution; even if the symptoms of a serious disease were readily recognizable in the early stages, few would willingly sacrifice an otherwise strong stock, and when a colony has become seriously affected, it is probable that the contagion has already spread.

PREVENTIVE MEASURES

For these reasons, it seems likely that whatever may in future be discovered, preventive measures by good sanitation and the cultivation only of vigorous stock will continue to be the chief bulwarks against apiarian disease. All who breed bees should cultivate only hardy stock: even in the presence of disease, rapid breeding often counteracts the loss to a great extent. To requeen stocks every year is perhaps a counsel of perfection, but it is certain that it is always poor policy to retain a queen who does not keep up a strong population.

One of the consequences of the introduction of movable combs was a tendency to retain them in use for many years, whereas under the old system combs were renewed almost every year. Systematic renewal of combs should be practised by the simple plan of providing each brood-nest with at least one new frame of foundation immediately after the honey-flow and removing in spring all those which are manifestly defective.

Hives should be subject to regular inspection and antiseptic cleansing to prevent the accumulation of waste matter. Healthy bees always remove rubbish from the combs, but much waste matter remains on the floor and forms a breeding ground for parasites if disease is present.

Attention should be paid to hive construction to make it difficult for apiarian enemies to enter. Mice are almost certainly disease carriers, and wasps undoubtedly are; every effort should be made to keep these pests in subjection.

The most dangerous disease spreader is the robber bee. All bees endeavour, when honey is scarce, to gather sweets wherever they can and raids are made on stocks unable to defend themselves. It is manifest that such stocks are those most likely to contain disease and this is the reason why the first colony to show serious disease is often the strongest in the apiary. Having overcome a weak stock, it carries home, not only its food, but the germs it contains and may be speedily and thoroughly infected. Do everything possible to prevent robbing by keeping all colonies up to full strength, reducing the entrances of weak stocks and taking care, when feeding not to spill syrup and never to feed until bees have ceased flying for the day.

A glass sheet placed over the entrance of a hive to confuse robber bees

Food is also very important. Artificial feeding is necessary in every apiary at times, but there is no surer means of introducing disease than by feeding with contaminated food. Combs are often taken from hives which have abundance and given to those which have not and, unless there is good reason to suppose these combs to be healthy, it is not without danger. Foreign honey or honey from apiaries unknown to the bee-keeper should never be given to bees. Foul brood has been introduced to many apiaries in this way. Sugar feeding, though sometimes criticized as unnatural, is much safer than feeding with honey not gathered by the bees of the hive, but raw sugars do contain impurities and may cause dysentery.

Water is also a source of possible contamination. It is difficult to prevent bees from drinking dirty water, but it is a help to provide suitable water fountains close to the hives.

Sound methods of bee-rearing, maintenance of colonies at full strength and vigilant sanitary measures will generally keep bee disease at bay.

ADULT BEE DISEASES

Dysentery

Under normal conditions bees do not discharge their excreta in the hive, but retain their food waste until conditions are suitable for a 'cleansing flight', which takes place, even in winter, whenever the temperature is high (over 50 °F.) and the weather fine. After one of these flights, excreta may be found on the outside of the hive and on other objects in the vicinity. This is a normal and satisfactory state.

Dysentery is characterized by the voiding of fæces inside the hive, the combs, walls and bees themselves being smeared with a brown muddy-looking fluid, with an offensive smell: the abdomen of many bees is swollen. The stock becomes sluggish and dwindles rapidly until, if the trouble is not checked, it dies out.

Some care is needed to distinguish between simple dysentery and the malignant type which is rather a symptom of serious disease than a complaint in itself. The simple form occurs only in winter and early spring, but dysentery may appear at other times, and in summer or autumn should be looked upon with suspicion.

In simple dysentery, the trembling of wings and aimless running about generally noticed in malignant diseases are not seen, although it is possible for the two forms to occur together. The best distinguishing feature is the nature of the fæces, which in the malignant form is solid and granular, instead of fluid.

Winter dysentery is chiefly due to unsuitable food and is often caused by the ignorance or carelessness of the bee-keeper, though the circumstances are not always under his control. Feeding impure sugar, or feeding so late in autumn that the syrup is not properly evaporated and sealed, are common causes. Damp, ill-ventilated hives have much the same effect, owing to the tendency of honey to take up moisture from the air. Some varieties of honey also cause dysentery, and evidence that heather honey is sometimes responsible has been gained at the West of Scotland Agricultural College. 'Honey-dew', which in some seasons is stored in large quantities, is well known to have this effect.

Prevention is always better and easier than cure. All stocks should be provided with winter stores not later than the end of September, so that they may be sealed and preserved in good shape. If feeding is imperative after then, soft candy should be given over the combs. Nothing but honey and refined sugar must be used for feeding. If impure food is given, the accumulation of waste in the bowel is so great that bees are compelled to void it in the hive.

A stock suffering from dysentery should be transferred into a clean dry hive as soon as conditions permit. If the combs contain fermenting food they must be removed and combs of sealed honey given, or candy supplied. Feeding with warm thymolized syrup is said to check the complaint. Dissolve 16 grains of thymol in a quart of hot water by shaking vigorously and mix 5 oz of this solution with the syrup made from 10 lb sugar.

Acarine Disease

This is the most prevalent of serious diseases affecting adult bees in Britain. Quarantine regulations have so far prevented its appearance in North America. It is caused by a species of mite, *Acarapis woodi*, which may invade the bodies of workers, drones or queen, to lodge itself in the thoracic trachea or breathing tubes. Here the whole process of reproduction takes place, generation after generation being produced till the tubes become congested and destroyed.

The disease is spread from bee to bee by female mites which leave and enter through the spiracles and it has been found that only young bees, less than a week old, can be invaded by the migrants, so that the trouble does not spread when there is no breeding, though affected bees become worse as the mite breeds in the trachea. The rate of growth and spread of infestation therefore varies and sometimes colonies known to be infested show no outward signs for months, and then suddenly begins the 'mass crawling' which precedes a speedy end. Externally, there are few signs in the early stages.

Most frequently in late winter, when a fine day comes after the bees have been confined to the hive a long time, crawlers are seen in front of the hive, running aimlessly about, climbing up grass stems and falling down when they attempt to fly. Many of them have one or both hind wings sticking out at right angles from the body. In an advanced stage, there may be thousands of crawlers and the stock soon dies out.

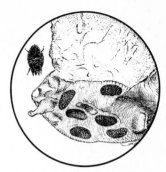

Trachea with the mites which cause acarine disease

Acarine disease may be present in one hive without spreading to others in the apiary, for the mites cannot live long outside the bee and can only be transferred to young bees. It is generally considered that the most prolific cause of spreading is 'drifting', that is to say, the tendency of bees, especially young ones, to enter a hive other than their own, when they have joined a larger 'playflight', or when hives are close together. It is also notable that when hives are in a long row, those at the leeward end become stronger because of this tendency to drift in that direction. Drones, in particular, are constantly changing home.

Swarming is clearly a very likely means of transfer, since when a swarm issues, any bees on the wing may join it, irrespective of the hive from which they came.

It does not seem likely that robber bees will carry this disease. They are usually old and incapable of picking up the mites and they hardly stay long enough in the hive to do so, but it is clear that if the method of checking a robber hive by transposing robbed and robbers is used, it is extremely likely that, if one is affected with acarine, the other will contract it in time.

Apart from these means of dissemination within the apiary, acarine can as readily be imported in swarms, stocks and nuclei as any other disease.

This disease can be readily diagnosed by anyone with a good hand lens, or a dissecting microscope. The technique perfected by the late Dr F. Thompson is as follows. The freshly killed bee is laid on its back on a cork slip, to which it is fastened by a needle driven between the second and third pairs of legs. The head and fore legs are cut off, leaving only a chitinal ring covering the affected part. By inserting a needle under this ring it can be removed and the trachea exposed. In a healthy bee, the trachea, recognizable as a spiral tube, is pearly white, but in acarine infestation it is stained with brownish patches. Those who have no skill in dissection can have the samples examined by one of the authorities mentioned on page 160.

THE FROW TREATMENT

If acarine disease is confirmed, treatment should begin as soon as possible. Several different methods are used, each with its own advantages and disadvantages. The one generally followed and considered the most effectual is that invented by Mr R. W. Frow and known by his name. The mixture is of safrol oil 1 part, Nitrobenzene 2 parts and gasoline 2 parts by volume. This is poisonous and inflammable and should be handled accordingly. The usual method is to sprinkle a prescribed quantity on a pad of felt about 4 in square and push this into the hive entrance, so that it rests in the middle of the floor. If the pad is attached to a stiff wire, it is easily drawn in and out. Some prefer to put the pad over the feed-hole in the cover, but either plan is satisfactory.

The amount to be used for a complete course of treatment is not more than 4·5 cc. of the mixture, 1·8 cc., exactly measured, being put on the pad in cold weather every alternate day for seven days. After the last dose, the pad is left for ten days and then removed. A prepared mixture may also be obtainable in special ampoules.

Several precautions are necessary. It is undesirable to use the remedy while bees are gathering honey, unless supers are first removed, but if treatment is deferred until after the honey-flow, colonies are liable to be robbed, for the odour seems to attract them. It is best to treat stocks in November when flying has ceased, and a piece of perforated zinc should be put over the entrance and removed in the evening, after bees have stopped flying, so that the imprisoned bees may take a cleansing flight. Plenty of air must circulate in the hive and when colonies are strong, it is a good plan to put a shallow rack under the brood-chamber before the treatment begins.

The mixture produces considerable mortality among the bees, but as these are probably the worst infested, this is of no consequence. Each day the dead bees should be raked out with a hooked wire and if they are very numerous, it is better to defer the next dose for a day. On no account should the quantity given be exceeded, or the whole stock may be killed. Some contend that it is simpler and more effective to give a full dose (4·5 cc.) at once, covering the entrance with perforated zinc for the first forty-eight hours, but it is best then to defer treatment till late winter after the bees have had a good cleansing flight.

Although there is abundant evidence of the effectiveness of this treatment, the readiness with which it causes robbing and the fact that a good many bees are killed, are serious drawbacks and some have given it up in favour of other methods.

ACARICIDE SMOKE

The latest treatment for acarine disease is an acaricide smoke, sold under the trade name of Folbex, which comes in the form of paper strips. Instructions are given with every purchase, but the procedure is simply to burn the paper strips in the hive. One dose is sufficient, after which the hive should be shut up for the night.

Nosema

This disease is caused by a protozoal parasite, *Nosema apis*, which attacks the wall of the bee's stomach. There it grows and multiplies rapidly and then passes into a spore stage, in which it is inactive, and is passed out with the excreta.

Being a disease of the digestive organs, it is naturally associated with the food. The spores are voided with the excreta, so that the surroundings of an affected stock form a prolific propagating ground. It has now been established that the disease is transmitted when healthy bees clean soiled combs. For this reason the changing of combs is a useful part of treatment.

Few signs are noticed in the early stages, but affected stocks do not progress. In slight infection, losses are small but if the disease gains ground the whole colony will dwindle and ultimately die out. There is also a stress factor involved. If bees have been shut up in their hives, especially when they are being transported, the disease will gain a more serious hold.

In recent years an antibiotic treatment has been found to be very effective. This is sold under the trade name of Fumidil-B which can be fed to the bees mixed with two gallons of sugar syrup either in autumn or spring. Coupled with a complete change of combs, this treatment is often entirely successful.

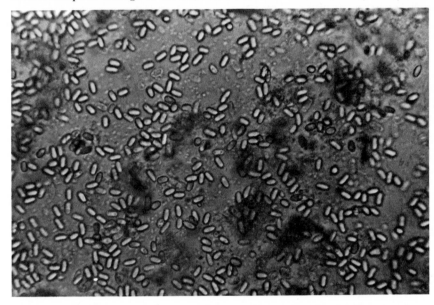

Nosema spores

In a serious outbreak, the most economical method in the long run is to destroy the affected colony, burning the bees and interior fittings and thoroughly disinfecting the hive. Nosema is troublesome in many parts of the Continent, and is a major scourge in America and Britain.

Paralysis and viral infections

Under this general name several rather different complaints have been described, and are now attributed to a virus and associated conditions and, since all kinds of disease are apt to show varying symptoms in different victims, they may all be due to one cause. It is generally recognized in bees sitting on the alighting board, shaking their wings and trembling in a palsied manner. If the brood-chamber is lifted, more will be found, often being pushed towards the entrance by other bees. Some of the victims are black and shiny, all their hair having come off, while now and then one will be found in flight, without apparent power to guide its movements, so that it drops aimlessly to the ground.

In my experience this occurs only in spring and, although there is considerable mortality and the stock never attains profitable strength, the signs disappear about July and all seems to be well. Indeed, I have had a stock suffering badly one season recover completely and do very well indeed the next year.

Some think this trouble is due to insufficient nitrogen and that the best treatment is to supply combs well filled with pollen. Others have claimed that dusting with sulphur has proved effectual.

Nine different varieties of paralysis have been enumerated which may possibly be associated with viral infection: (1) Infections, in which the symptoms described above occur; (2) Genetical, in which it is said that newly emerging bees are often hairless and some have malformed wings; (3) Nitrogen deficiency, caused by lack of pollen in brood rearing; (4) Damaged pollen. In this form, the bees are not hairless, but have swollen abdomens. It is thought that injury by frost is responsible for this; (5) Poisoned pollen. There is no loss of hair, nor is the abdomen swollen, but trembling and inability to fly are noticeable. This is due, it is thought, to some doubtful species of plant such as rhododendron, scabious or foxglove; (6) Poisonous nectar; (7) Poisonous honeydew; (8) Fungal poisoning; (9) Arsenical poisoning.

The similarity of symptoms may make it difficult to decide on the causes of outbreaks of paralysis. It may be that there is an infectious type which needs to be treated seriously, but for the most part these complaints are but passing and disappear when the cause, whatever it be, is removed. Some strains of bee seem more susceptible to this than others, and one practical treatment is to requeen the stock with a queen of a different strain.

Amœba Disease

This is a spring complaint fairly common in Britain, though difficult to detect, and frequent in Europe. Bees are found dying round the hive and dwindling continues for some time. It is sometimes associated with nosema and responds to the same treatment. It is not common in America.

The trouble is caused by an amœbal parasite affecting the Malpighian tubules. It is transmitted through the excreta. Bees do not seem to lose the power of flight at an early stage as they do in acarine disease, nor are they found crawling in the masses associated with advanced acarine infestation. Bees leave the hive slowly and fall weakly on the ground with their legs shaking.

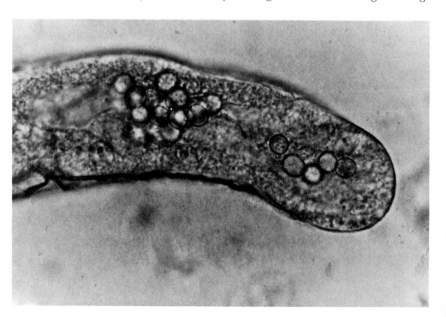

Amœba cysts in the Malpighian tubule

Freaks and Deformities

Abnormal bees are sometimes found in colonies otherwise in good health. Some such freaks appear to be half male and half female, while wings, legs, or other organs are absent in others. None of these abnormalities need cause anxiety, as such freak bees are incapable of mating.

BROOD DISEASES

Foul Brood is the most troublesome bee malady and responsible for very heavy losses. The name seems first to have been applied to the disease by Schirach (1769) and he also succeeded in effecting a cure by starvation, not unlike that sometimes practised now. The existence of more than one variety was recognized by Dzierzon and Dr C. C. Miller, before bacteriologists showed that there was a difference between the cause of the 'ropy' and 'stinking' types.

American Foul Brood

This is due to *Bacillus larvae*, a slender rod, which form long chains that develop into innumerable spores. Sometimes corkscrew-shaped objects (whips) are present, but not always. The spores are minute ovals without appendages. The dead brood is generally sealed or about to be, but the bees often uncap some of the cells. The affected larva is coffee coloured as long as it is moist, but finally dries up into a rough dark brown scale, which adheres firmly to the cell wall so that neither the bees nor the bee-keeper can detach it. The moist decayed brood is 'ropy', drawing out, if stretched, in strings as much as an inch long.

Bacilli multiply at an enormous rate by constant division of the rods, this process having being seen to occur twice within an hour. When, owing to destruction of the larva and consequent consumption of its food, the active life of the bacillus ceases, a spore is formed and this remains inactive till brought into contact with favourable conditions. It is by these spores that the

American foul brood

disease is spread. Robber bees attack a stock weak from foul brood, carry spores to their own hive and supply them to their brood. Or the bee-keeper may transmit them by manipulations, or by feeding honey from one stock to another. Swarms from stocks with foul brood may also carry spores with them.

In the British Isles, American foul brood is by far the most frequent. In 1948 nearly 2,000 stocks were found affected with it. A regular system of inspection is now carried out by appointed officers and it is hoped that in time it may be brought under control.

In the early stages of American foul brood there is no outward sign and a beginner may fail to notice it even when the combs are examined, because in this disease the larvæ rarely, if ever, die before they are sealed over. If a comb of capped brood has irregular holes in some of the cappings, those cells should be opened and the contents examined. A good comb of sealed brood should be plump-looking, the cappings slightly convex, but if there are any here and there which are sunken and darker than the surrounding ones these also should be opened. If a toothpick is pushed in and comes out with a sticky brown mass adhering and drawing out in a thread, American foul brood is to be feared and prompt steps must be taken to deal with the matter for, if neglected, it will assuredly become worse and, as soon as the stock becomes weakened, robbers will carry the spores into their hives.

Beyond doubt, the wisest step to take is to destroy a stock clearly affected, but unless the case has so far advanced that it is obvious to the most inexperienced, a report or sample should be sent, either to the State apiary inspector or to the US Department of Agriculture (see page 160).

DESTROYING THE STOCK

If it is decided to destroy the stock, the simplest method is to close the entrance in the evening after all bees have come home. Seal it up with clay or moistened earth. Through the feed-hole in the cover or quilts pour in about half a pint of gasoline and cover up with sacks so that fumes cannot escape. The bees will be dead in a very short time, but meantime a hole should be dug about 2 or 3 ft across and of the same depth. Some inflammable material is placed in the bottom and all the contents of the hive—bees, combs, brood and cover—arranged on this and set alight. If the hive is old or badly constructed, this can go on as well, but there is no need to destroy a sound hive. When all is reduced to ashes, it should be well buried with soil.

Supers of honey must be destroyed, and the greatest care must be taken in their destruction to see that bees cannot get access to the honey which should not be consumed at home. On no account must any risk be taken of its being fed to bees or becoming accessible to them. After extraction the combs should be destroyed. The hive should be disinfected by scorching with a blow lamp, or if this is not available it can be painted inside and out with creosote and thoroughly dried before being used again. All tools used in these operations must also be thoroughly cleansed and disinfected—metal ones must be boiled in strong soda water and wooden articles be painted well with creosote. Bee-keepers who are members of associations can insure themselves against loss.

It is rarely advisable for a novice to attempt any treatment of foul brood, but mild cases can be successfully cured providing proper precautions are taken. The essence of treatment is the complete removal of all food in the hive and the destruction of the brood. The bees are shaken into a box and kept confined for about twenty-four hours, by which time they will have consumed all the food in their bodies. During shaking, care must be taken not to spill honey about and it is best to do it over newspaper which can afterwards be burnt with the brood combs. After their confinement, the bees are hived on new frames of foundation and fed to build them up.

SULFA DRUGS

During the last few years, many people have been experimenting with sulfa drugs and there does not appear to be the least doubt that sulfathiazole, mixed with the food, arrests the progress of the disease, but it has no effect on the spores and the complaint generally breaks out again if old stores are left in the hive. On this account the authorities have not given their blessing to this treatment, but by removing all sealed stores before the treatment is given, every particle of food in the hive will be given a dose of the drug, since bees are constantly shifting unsealed stores.

The simplest method of treatment is to dissolve 1 gram of sulfathiazole in 1 pint of thin syrup and supply it in the usual feeder. After the first dose, $\frac{1}{2}$ pint is given every other day for about three weeks, or until symptoms disappear. It is also available in tablet and powder form.

Mr G. A. Taylor, who has had much success in treating this disease experimentally, says that results are much quicker if the drug is sprayed into the combs. He uses 1 gram sulfathiazole to $\frac{1}{2}$ pint of spirit. The bees are shaken from the combs, which are then sprayed thoroughly with the solution. This is followed by a feed of syrup, containing 1 gram of sulfathiazole to every gallon, *to every stock in the apiary*.

European Foul Brood

This is generally considered to come from *Bacillus pluton*, a short spindle-shaped rod, which forms groups, not chains, without forming spores. Unlike *B. larvæ*, it has not been proved by direct feeding experiments to cause the disease, but two interesting facts are known. Lochhead, in Canada, has shown that there is some reason to suspect that *B. pluton* may be a stage in the life cycle of *Bacillus alvei*, long known to be associated with the disease; and Fyg, in Switzerland, has found rod-like organisms in the tissues round the stomach of sick larvæ, *B. pluton* being present in the stomach.

The names applied to these diseases are rather misleading, for both occur in Europe and America. The name 'American' was given when American research workers demonstrated the difference between the two diseases, which had hitherto been regarded as slight variations of the same one.

European foul brood differs from the other chiefly in that the death of the larva occurs soon after it is weaned, about the fourth day from hatching. It lies often in a twisted position, instead of neatly curled in the cell, turns yellow, collapses and gradually dries up into a loose brown scale. Often it has a vile odour.

In Switzerland, where both diseases have been studied under Government supervision for many years, radical difference between the two diseases, both as to occurrence and treatment has been found, for while American foul brood occurs mostly in good seasons, European is closely associated with bad years and the conclusion is being reached that the latter does not get a hold unless the bees are in poor condition. Direct experiment has shown that a strong

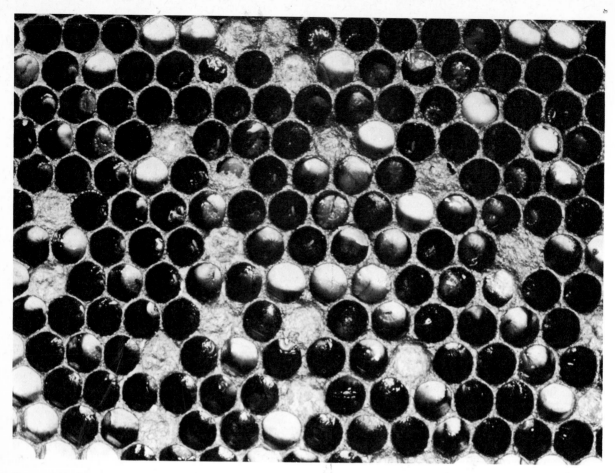

European foul brood

colony with plenty of stores does not contract European foul brood even when supplied with diseased brood.

Chalk Brood

A complaint which occurs fairly frequently, but is generally confined to a few patches of brood. It is caused by a mould related to the common pollen mould, which turns the larvæ into tough mummified corpses. As they dry up they become loose enough to shake out of the cells. It seldom occurs to a serious extent, but it is advisable to take an early opportunity of removing combs in which the complaint has manifested itself.

Stone Brood, Sac Brood and Addled Brood

These are other mild forms of disease differing slightly in their symptoms. They are not common in Britain and are not virulent in their effects or seriously contagious.

Sac brood, however, is fairly prevalent in the United States. It is a virus that affects the larvae at the end of winter, spring and early summer, and is associated with the other stress diseases, European foul brood and nosema. At present no drugs have proved to be wholly effective, so preventive measures

158

are necessary. Steps should be taken to ensure that the hive is raised above damp ground and that a clean water supply is available to the bees.

Chilled Brood

This sometimes scares the beginner into the fear of foul brood, but is very different in its cause and appearance. The larvæ turn grey and gradually blacken, while the bees remove the corpses, which they rarely attempt to do in foul brood. Chilling may occur owing to rapid weakening of the colony from acarine, or poisoning, but as a rule the balance between workers and brood is properly adjusted so that all brood can be covered safely in a cold snap.

Untimely manipulation, especially too early 'spreading brood' is more often responsible, while formation of nuclei with insufficient young bees, may result in most if not all the brood being chilled.

In the summer time, during a long spell of cold or wet weather, the bee-keeper may be disturbed to see brood lying in front of the hive. This is a normal feature of hive life, but it indicates a shortage of food. Almost invariably the larvæ to be thrown out are drones, since these are the first to be sacrificed if starvation impends, but if the shortage becomes more severe, worker larvæ may follow. The remedy is to give a good supply of food.

Summary of Brood Diseases

Name	Cause	Signs
American foul brood	*Bacillus larvæ*	Brood sealed, coffee coloured, ropy, drying to rough scale adhering to lower side of cell wall.
European foul brood	*Streptococcus pluton*	Larvæ young, unsealed, yellow gradually darkening to black. Scales easily removed.
Sac brood	*Filtrable virus*	Larvæ sealed, yellow to black, outer skin rough, contents watery. Sac easily removed when dry.
Chalk brood	*Pericystis apis*	Larvæ hard, whitish with dark spots.
Chilled brood	*Accidental*	Larvæ grey to black; after death thrown out of hive.

At all times, and particularly when disease is known to be in the neighbourhood, great care should be taken in manipulating stocks. Appliances used for a suspected stock should be disinfected before being used for another, as should the hands of the operator. In dealing with known cases of foul brood, overalls should be used to prevent the clothes being soiled and changed for a clean pair when healthy stocks are examined. Combs from infected or suspected stocks should on no account be given to healthy ones, but destroyed.

Every bee-keeper in the United States has a free service available in respect of adult bee diseases, and it is in everyone's interest that regular health checks be made. For this purpose, it is necessary that samples of bees should be sent to one of the appropriate places mentioned on page 160.

The symptoms of adult bee diseases may be superficially similar, whether the cause be acarine, nosema, paralysis or even poisoning; an excessive amount of dead or crawling bees in front of the hive, or perhaps a colony just dwindling away.

To diagnose the cause visually is frequently difficult, even for an experienced bee-keeper, and microscopical confirmation of on-the-spot diagnoses is always desirable. Since each disease or condition requires different treatment, it is important that an accurate diagnosis be made.

The best possible time for the assessment of most adult diseases is during the period March–April but routine examination at any time during early spring will normally reveal trouble if it is there. In definitely suspected cases, take a sample at any time rather than let it continue unchecked.

Samples of bees, dead or alive, should be packed in porous containers of wood or cardboard; glass or plastics should be avoided because decomposition in these is often so rapid that the bees arrive in a condition which makes them quite unfit to handle. Matchboxes make the ideal containers; these and similar boxes are recommended whenever possible. When packing bees for the post the following points should be remembered:

1 A sample should contain 25–30 bees (a full matchbox holds about 50).
2 Food should not be put in with the bees.
3 The container should be clearly marked with the colony identification for easy reference.
4 If the inner tray of the box seems liable to slide out, secure it with a pin through the side.
5 The sender's name should be legible (this is an obvious point but frequently overlooked).
6 Relevant information which might assist diagnosis should be given in a covering letter.
7 A letter should not form part of the wrapping of the box.
8 'Bees' or 'Live Bees' marked plainly on the outside will often hasten delivery.

Since it takes two to three weeks for evidence of some diseases to develop with individual bees, the sample should consist of older bees. Those on the outside of the cluster are likely to be older than those on the face of a brood comb.

If there are recently dead or crawling bees in front of the hive, it is among these that disease, if any, is likely to be found. There is no need to disturb the colony; pick up about thirty of the freshest bees, put them into a matchbox and prepare for posting. Freshly dead bees may also be found on the floor-board in the spring by gently raising the brood box and placing it aside.

If there are no dead bees or crawlers available, then one of the simplest and quickest methods of getting a live sample is to scoop the bees into an inverted matchbox tray from a flat surface: the front of the hive, underside of the crownboard, or the face of an outside comb.

Where there is a State apiary inspector, and most of the States do employ them, he is the person to whom samples should be sent. Samples of suspected brood disease can also be sent to the Bee Pathology Laboratory, Entomology Building A, Agricultural Research Centre, Beltsville, MD. 20705, where a diagnosis is made free of charge.

ENEMIES OF BEES

It is only natural that a creature which collects large stores of honey should have enemies, for this is a food highly esteemed throughout the animal world. Against most of them the defence provided by nature—the sting—is adequate, but there are several enemies against which the bee can only hold its own by constant vigilance and it is important that the bee-keeper should know what to do to assist in defence against these attacks.

Wasps (yellowjackets)

Wasps are the commonest and most troublesome. *Vespa vulgaris*, the common wasp, *Vespa germanica* the German wasp, and *V. crabro* the hornet, are the most plentiful, but the latter is not very much trouble in Britain, for its colonies seldom attain great size. In some parts of Africa and the Mediterranean, however, it is a serious nuisance.

Once in a while I have seen a queen wasp enter a hive in spring, but generally speaking wasps do not trouble the apiary till late summer; from August to October they are a constant menace. At first they hover over the ground in front of a hive and pounce on home-coming foragers who are heavily laden and miss the hive entrance. The wasp pounces on the bee's back and severs the body above the abdomen, which it carries away. This goes on continually and is not very serious, but the easier they find the task, the more wasps come, and if there is not much resistance they begin to enter the hive. They are on the wing much later in the day than bees and are often able to slip inside when the guards have withdrawn.

Strong colonies can deal with wasps, for guards attack them on sight, but weaker stocks, such as queenless ones and nuclei, can soon be overcome, and the honey and brood carried away. The bee-keeper must keep his eyes open and assist such stocks by every possible means, reducing the entrance to one bee-space. Care must be taken not to spill honey or syrup about or it will invariably attract wasps. They have an uncanny knack of getting into hive roofs and taking syrup from the feeders and any possible back door into the hive will certainly be discovered by them.

The honey house must be bee- and wasp-proof at extracting time, or wasps will be an intolerable nuisance. They find their way through places a bee would not discover and too much care cannot be taken to stop up cracks and holes. Wasps' nests in the vicinity must be hunted out and destroyed, the simplest plan being to push a piece of cyanide of potassium into the entrance as far as possible, pour water after it and stop up with a thick turf. The moistened cyanide gives off continuous fumes which permeate the nest and kill its occupants.

Bottles containing some fermenting liquor—jam scum is excellent—can also be used for trapping wasps. A zinc cone forming an entrance into the bottle will ensure that, once in, they do not get out again. So long as the liquor is acid, bees will not enter the bottles.

Wax Moths

Several members of the family *Gallerideæ* inhabit the nests of bees and two of them are entirely associated with the honeybee. The largest is *Galleria mellonella L*, the greater wax moth. It is about an inch across the wings, which are grey, the upper ones with purplish black markings. These moths are on the wing during most of the summer, July being the time of greatest abundance. On calm evenings they fly round the hives and sooner or later enter, to lay their eggs in the combs. When the caterpillar hatches, it burrows into the comb, consuming the wax as it proceeds and protecting itself by lining its tunnel with silk. It is greyish and semi-transparent and can dodge quickly to and fro in the tunnel. It takes about three weeks to reach full size and then settles in some place more or less out of the way of the bees, such as the saw-cut in the top of a frame—a favourite spot when cloth quilts are used—or behind a dummy board. There it makes a strong silk cocoon and in due course emerges as a moth.

This is a southern species and rather local. It is more destructive in the southern states than in the northern states and considerable care is needed to keep it down. Strong stocks do not suffer seriously, but I have seen poor ones turned into a mass of silk webs with not a scrap of wax left in the frames. Periodical cleansing, destruction of moths and larvæ wherever seen, and renewal of infested combs will enable you to cope with this pest.

The small wax moth, *Achroeia grisella* Fab., is little more than half the size of the other species, but it is far more widespread and abundant. It is silvery grey without markings and the female is considerably the larger. It operates in much the same way as the greater wax moth, but it is particularly troublesome in combs stored away from the hives. Many a bee-keeper who has not inspected his spare combs for a season has been disgusted to find them reduced to skeletons. Stored combs should be kept in moth-proof racks, the best plan being to wrap the racks in newspaper immediately they are removed from the hives. If there is any doubt whether moths have already laid their eggs in them, they should be fumigated. The simplest and safest method is to pile the supers full of combs in a stack, paste paper round the joins between the boxes and put an ounce or so of paradichlorobenzene (P.D.B.) crystals on top of the pile, covering it over with sacks to keep in the fumes. As this evaporates very slowly, it kills the larvæ as they hatch from the eggs.

Bee Louse

This insect, sometimes called blind louse and scientifically *Braula coeca*, is a

Wax moths
Left: *Galleria mellonella* L.
Right: *Achroeia grisella* Fab.

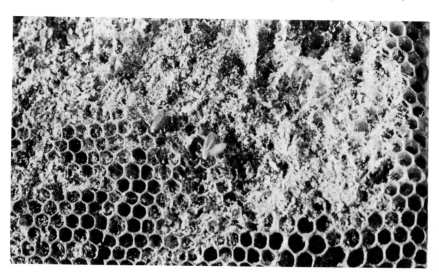

Wax-moth damage to comb

fly of the dipterous order, parasitic on bees. It is about one-eighteenth of an inch long, reddish brown and in general appearance like a minute spider. Its head is large and is remarkable for the absence of organs of vision. The abdomen is large and round and the legs have claws which enable it to cling to the hairy body of the bee. Compared with its host, this parasite is exceptionally large, and, although workers and drones rarely have more than one on them, the queen is particularly susceptible, over one hundred having been taken from a single queen.

Like all insects, these lice have four life stages. The eggs are laid on the cappings of comb and the larvæ tunnel along just under the surface, pupating in due course at the end of the tunnel. When mature, the perfect insect attaches itself to a bee, usually on the upper side of the thorax and lives by sucking honey from the tongue of its host.

This is a southern insect and abundant in the south of Europe, but it is not found everywhere in Britain, being little known in the north and only regularly in the southern counties.

Apart from the damage to comb cappings, these creatures do not do much harm. They are very sensitive to tobacco smoke, which makes them let go their hold and drop to the floor. If they are found to be numerous, it is worth while to blow tobacco smoke into the hive, remove the floor immediately and clean it off with the blow lamp.

Ants are fond of honey and bees do not seem to mind them. They crawl up the hive legs and pilfer small quantities. It is difficult to keep them out, but standing the legs in tins of paraffin, or creosote or tar will serve, until the stuff evaporates. If the hives stand on rails, sacking soaked in the stuff can be wrapped round them between the hives.

Many different kinds of spiders are found in and about hives, but though they may take an occasional bee, they are not of sufficient consequence to bother about.

Birds

Certain insect-eating birds, notably tits and flycatchers, take toll of bees. I have often seen a titmouse perch on a stump near a hive during winter, dart down and tap on the flight board, snap up the bee which came to investigate and repeat the process several times. One summer afternoon I watched a pair of flycatchers coming and going to a hive for twenty minutes, during which time they must have taken fifty or more bees. The method adopted was to alight on top of the hive, dart into the crowd of homing bees and fly off with a victim. It would not have been any use to put netting over the flight board, an effective plan against other birds, so I stuck a bamboo cane on each side of the hive, projecting forward 45°. Between these canes I fixed black cotton in lines about 4 in apart. I never saw the birds again, either at that hive or any other.

Badgers

In some districts, it is not unknown for badgers to overturn hives and consume the contents. The only thing to do in such places is to make sure the hives are securely fenced in, remembering that badgers can dig under a fence not sunk into the ground.

Mice

I have already mentioned the precautions to be taken against mice at the beginning of this chapter, which are found everywhere and will enter hives for the winter if they can.

Toads

Toads sometimes take up their stance near a hive and feed on bees falling to the ground, but perhaps they do as much good as harm by getting rid of sickly bees which may convey contagion.

Bears

In the mountainous and forested areas of the United States bears do present a problem and can cause extensive damage to the apiary. Bears eat bees, brood, and honey and thereby destroy the colony in the process. It is difficult to prevent them from attacking an apiary, since electric fences are too costly for most bee-keepers, but they are less likely to raid a hive that is positioned close to a house or barn.

Skunks

Skunks present a serious threat to the colony's safety in many parts of the country. They have a tremendous appetite for bees and can severely weaken a colony by their marauding tactics. Fencing or raising the hive out of harm's way can help in dealing with these animals.

Squirrels

Red squirrels, like mice, often build their nests for the winter months in the hive or honey house where they can destroy many combs in their search for food.

16
Products
of the Hive

HONEY

As found in the hive, honey is a mixture of various kinds of sugar with water, flavoured by floral essences and containing traces of various minerals, nitrogenous substances and an enzyme (invertase). It differs from the nectar of the flowers from which it is derived mainly in the reduced water content and the transformation of certain kinds of sugar (sucrose) into other kinds (dextrose and lævulose). There is great variety in the composition of the original nectar, not merely in different species of flowers, but to some extent in the same kind of flower, according to the conditions of soil and climate. Some flowers give mainly invert sugar, others chiefly sucrose, but most of the latter is inverted in the body of the bee before it is put in the honeycomb, so that the two sugars which predominate in pure honey are dextrose (glucose) and lævulose (fructose).

Not all flowers produce nectar, but by far the larger number of those with brightly coloured petals, as well as some of the less conspicuous kinds like the grape, secrete it in varying quantity. In some it is contained in clearly visible glands on the petals, in others it is partly concealed at the base of the petals, while in many it is ingeniously hidden by various contrivances, so that only certain insects with specially adapted organs can reach it. This ensures that the insect will not take away the prize without paying for it in the coin of cross pollination.

Broadly speaking, warmth is the main factor in regulating the amount of nectar secreted by flowers and there appear to be very few which yield it at a temperature lower than 50 °F. In summer time, yield seems to be plentiful at temperatures above 65 °F. and reaches its maximum with a day reading of 80 °F. Clovers, in particular, appear to need considerable warmth and it is notable that only when the thermometer is well into the eighties can honeybees obtain nectar from red clover, because it then rises high in the corolla tube: normally, their tongues are not long enough.

The attraction which certain flowers have for the bee also depends on the percentage of sugar their nectar contains, but it is seldom above 30 per cent and on the average the water content is between 70 and 80 per cent. There is also considerable variation in the time of day during which nectar is secreted freely, since some flowers yield it in the forenoon, others in the afternoon. But some, like borage, to judge by the constancy with which bees visit them, maintain an equally constant yield. There is even an occasion on record when bees visited basswood by moonlight on a very hot summer night. Attempts have been made to estimate the possible yield of honey per acre of various flowers, but this varies so widely that the results are clearly only speculative.

CONCENTRATION

In addition to inverting sucrose by ingesting nectar, bees reduce the amount of water by manipulation: that is, they pour it slowly into the cells from their honey stomachs and shift it from cell to cell until it reaches the stage of concentration at which they seal it over with a waxen cap. During a heavy honey-flow, numbers of fanning bees are stationed at top and bottom of the hive, in order to produce a continuous current of air that carries off the moisture, which can be seen condensing on the floor and even running out of the front. Sealed honey contains an approximate average of 17 per cent water.

Sugars are soluble in water, but dextrose is less so than the other two forms, and it is this sugar which is chiefly responsible for crystallization or granulation, as bee-keepers usually term it. Of itself, lævulose does not crystallize. Stirring or otherwise disturbing a thick sugar solution hastens crystallization, which accounts for the well-known fact that honey in the comb remains liquid longer than extracted honey.

The invertase in honey gradually changes (inverts) any sucrose there may be present, but heating above 140 °F. kills this enzyme, so that inversion ceases. The familiar phenomenon of 'frosting' which often happens when honey has been bottled without heating is probably due to concentrations of dextrose crystallizing more rapidly than that which has a larger proportion of lævulose.

The minute quantities of minerals, which cannot be too highly rated from the dietetic point of view, consist of iron, phosphorus, calcium, sodium, potassium, sulphur and manganese. There are also traces of essential oils, gums and other flower constituents, including grains of pollen. Though very minute, pollen grains are all different and characterize the various flowers, so that it is possible to ascertain the source of a sample of honey by examining the pollen grains in it. This is a highly skilled task, but can be done with such precision that the Courts accept evidence of pollen grains in honey in cases where the source of a sample is disputed. Thus if honey sold as British contains pollen from a flower not growing in the British Isles, it is obviously not pure British honey and convictions have been obtained in several cases of the kind.

SOURCES AND VARIETIES

Although it is probable that honey taken from a hive is never composed absolutely of one kind of flower nectar, there are certain times of the year and particular places where the income over a considerable period comes predominantly from one species. In wooded districts, for instance, the early honey comes principally from certain trees—willows, holly, crab-apple, maple, etc. In fruit-growing districts, gooseberries, plums, pears, apples and cherries are the basic sources. All these early trees combined produce a mixture which is rather dark in colour and rich in flavour. Later on the honey-flows are usually heavier and more concentrated on a single source. Where there is a field of sainfoin handy, for instance, all the honey will be bright yellow and mild in flavour. Where, as in so many places, white clover is the main source, much paler honey with even more delicate flavour will come in during July.

In heather districts two distinct forms are recognized. One is the almost pure product of ling, which, perhaps more than any other, is likely to be almost 100 per cent pure, giving the dark brown or purplish jelly-like honey unlike any other. The other is 'heather blend', in which the ling is more or less mixed with honey from bell heather, cross-leaved heath, clover and other late flowers. Such samples usually come from places on the edge of moorlands where the bees can range equally well over neighbouring clover fields or over the moorland.

Canada, New Zealand and the USA send out excellent honey and Jamaica also exports honey of good quality, except that, owing to local conditions, it contains a very high sucrose content, which is not entirely inverted by the bees. European countries mostly consume their own honey and several of them import more than they export. The chief objection to mass-produced honey, which is often equal to local produce, is that it is imported by large concerns, whose chief idea is to market a product of uniform appearance, each jar appearing exactly like all others. Since the natural colour of honey varies infinitely, this result can only be achieved by mixing large quantities of all sorts together to produce the attractive golden colour regarded by the man in the street as typical. Not only does this destroy the individual flavour of each sort, but since mixing can only be done by heating the imported granulated honey and stirring it in a liquid state, some of the floral essence is driven off during this process.

The bee-keeper who wishes to build up a steady demand for his honey must see that his produce is put on the market in an attractive form, without injuring its distinctive qualities. As I have already pointed out, the popular demand is for light mild-flavoured honey. This is due to the fact that such honey is commonest, and being transparent, sparkles in the light. It is possible to educate people to prefer dark, strong-flavoured honey, but the fact must be faced that it does not look so attractive in the jar as the clear, golden type. When it happens to be, in some years, almost black, it is very hard to sell.

One way of making things easier is to pack such dark honey in opaque containers, like the wax paper pots or handmade 'quaint' earthenware jars when they are available. They are most attractive containers and enjoy a certain popularity as gifts.

Each season's crop varies to some extent from others, but all is rarely the same colour, and it may be worth while, if there is only a small quantity of light honey, to keep it separate so that customers may have a choice. On the other hand, by mixing all together at the time of extraction the general tint is lightened.

Honey should be sold either clear or granulated. The stage between the two is always, even with light honey, rather repellent, so when granulation begins it is much better to wait and sell only those jars which have become quite solid.

The better the honey, the harder it sets, and here again I would prefer to teach my customers to appreciate that this is the real Simon Pure, but, as Mr Manley points out, the untutored public likes to be able to 'spoon it out easily'. This means that many, like Manley himself, 'cream' honey in this way. After having set hard, the honey in the tins is warmed either in a boiler of hot water, or in an electric heater sold for the purpose, until it is soft enough to turn out into the bottling tank, where it is stirred to a smooth cream and bottled.

167

USES OF HONEY

Natural foods like milk, fruit, green leaves and nuts are generally best when eaten raw and this is true of honey, which shares with milk the distinction of being intended by nature solely as a food. If we can convince the public of this, honey need not be used in any other way. In one respect its nature is actually superior to that of milk, for whereas milk is known to be a dangerous medium for the transmission of disease, honey is not merely completely safe in this respect, but it is actually anti-bacterial and beneficial in several common complaints.

The ease with which honey is sold in times of sugar shortage, may lead some to suppose that there is never any trouble in disposing of it, but there have been times when it was extremely difficult to do so, for the general public knows very little about food values and often pays more for articles which look attractive, but have much less real merit. Sometimes when several good honey seasons in succession glut the market, commercial bee-keepers have some difficulty in selling their crops. Fortunately, when properly packed, honey remains in good condition for years and when a lean season comes, the surplus soon disappears.

The commercial bee-keeper has his own organization for marketing, and it is the small bee-keeper who sometimes finds it difficult to dispose of his crop after a bountiful season. He ought, therefore, to know of the many uses to which honey can be put, not only for himself and his own household, but for his neighbours who are potential buyers.

Honey in the comb is always more readily saleable than extracted honey and commands a higher price. Those who have once eaten it realize that it is far superior in flavour because it retains unchanged all the delicate essences which the bees sealed up in it. It does not have to compete with the imported product, for its fragile nature makes it an expensive article to transport. If wrapped up as described in the section on packing and stored in a dark dry place, it will generally remain liquid for at least a year, if it should be necessary, which it rarely is, to keep it so long.

Production of sections is, however, a difficult art and most bee-keepers strive mainly for extracted honey, since the yield of this is always larger— as much as 25 per cent— and there is less risk of swarming. The essential thing when preparing extracted honey is to ensure that the natural flavour is preserved to the fullest extent. I am strongly opposed to any manipulations which pander to the public eye, for in the effort to secure a clear, uniformly golden appearance or a soft, creamy, granulated honey, much of the flavour is lost and the valuable invertase, most beneficial to digestion, is easily destroyed.

GRANULATED HONEY

Honey which has been heated granulates slowly if at all, and so keeps clear longer. It does not set hard in the jars or tins, but remains soft. In this form, it is easier to remove from the jar, which is why some people prefer it, but its flavour is always inferior.

Most connoisseurs prefer granulated honey, but those who like it liquid can always bring it back to this state by standing the jar in hot water. Clear honey is delicious on bread, bread and butter or toast and its slight laxative property renders it more desirable than marmalade as a breakfast sweet for many people. For sweetening porridge it is superior to sugar and it is especially acceptable with cereals.

Sandwich Filling

Beat together equal parts of butter and granulated honey until well mixed, but not too soft. Add chopped raisins or nuts.

Grapefruit and Honey

Put two or three teaspoonfuls of honey in the centre of a halved grapefruit, after removing the centre membranes.

Raspberries and Honey

Put hulled raspberries in a deep dish, spread cream or top of milk to cover, squeeze a few drops of lemon juice on top and drizzle honey over. Allow to stand a few minutes, then stir until the berries are coated.

Honey in Preserves

Honey can always be used instead of sugar in making jams, jellies and syrup for bottling, but certain points must be remembered. First, honey is sweeter than sugar and may make the preserve too cloying, so it is usually better to use half honey and half sugar. Honey also contains some water, making longer boiling necessary and it foams up when heated, so when used for jam making, it must be watched carefully at the start lest it boil over. Nor must it be forgotten that honey has its own flavour which does not always blend happily with the fruit. I consider it better to bottle fruit without syrup and add honey when it comes to table.

Sunshine Preserves

Mix a pound of honey with a pound of soft or stone fruit, spread out in shallow dishes and expose to strong sunshine till the mixture becomes thick. A good way to do this is to put the dish in the solar extractor used for wax (see page 176). A sunny greenhouse or garden frame can also be used but, of course, the dish must be quite safe from bees, wasps and flies. Finally pack in sterilized jars and seal.

Honey Plum Butter

Wash plums and cook gently in water until soft. Pass through a sieve and for each cup of pulp, add half a cup of honey. Cook slowly until thick and jelly-like. Pour into hot sterilized jars and seal.

COOKING

The uses of honey in cooking are not nearly so well known as they should be. Here is an economical recipe for

Steamed Honey Pudding

6 oz self-rising flour, 2 oz butter, 3 tablespoons honey, pinch of salt and enough milk to make a stiff dough.

The fat and honey can be rubbed into the flour, or melted and poured into a well in the flour. Mix with milk to form a stiff dough and steam in a greased basin, covered with fat, for $1\frac{1}{4}$ hours. This pudding can also be baked in a pie dish, or varied by putting a spoonful of jam at the bottom of the basin or dish.

Honey and Chocolate Mould

Mix 2 oz rice flour, 2 oz cocoa and $1\frac{1}{2}$ tablespoons corn starch to a smooth paste with cold milk. Bring $1\frac{1}{2}$ pints of milk to the boil, add to the paste and stir well. Simmer ten minutes and add $1\frac{1}{2}$ tablespoons honey and 1 oz butter. Simmer again 3 minutes and pour into a wetted mould to set.

Honey Pancakes

4 oz flour, $\frac{1}{2}$ oz superfine sugar, $1\frac{1}{2}$ tablespoons honey, 1 saltspoon salt, $\frac{1}{2}$ saltspoon nutmeg, 2 eggs, $1\frac{1}{2}$ cups milk. Mix all the dry ingredients and stir in the honey. Beat up eggs and mix with milk. Pour gradually into the mixture and beat until smooth and thick. Fry in the usual way.

CAKE MAKING

In cake making, honey has definite advantages, but it must be used with judgment or it is not economical. Cakes made with honey keep moist longer, but if too much is used or it is not well blended, they can become too heavy. Ideally, they should be free from other flavouring, but fruit and spices are commonly used.

Plain Honey Cake

Beat well together $1\frac{1}{4}$ cups sour milk, 6 oz sugar and 4 oz honey. Work this well into 10 oz flour. Bake in buttered pans from a half to three-quarters of an hour and serve hot.

Honey Scones

$\frac{1}{2}$ lb self-rising flour, 1 oz butter, 1 oz honey, 2 oz dried fruit, pinch of salt. Mix flour and salt and rub in fat. Add fruit and honey and mix with milk to a light dough. Roll out and cut into rounds and bake in a quick oven.

Honey Gems

Mix 6 oz honey in $2\frac{1}{2}$ cups sour milk, work in enough flour to make a soft dough and bake in heated tins.

Honey Teacake

1 cupful honey, $\frac{1}{2}$ cupful sour cream, 2 eggs, $\frac{1}{2}$ cupful butter, 2 cupfuls self-rising flour. Mix and bake as for preceding recipe.

Honey Sponge

1 cupful honey, 1 small cupful self-rising flour, 5 eggs. Beat yolks and honey together and whip whites until stiff. Carefully fold the yolk mixture into the egg whites. Then fold in the flour, without beating, adding lemon or other flavouring. Pour into a greased, floured pan and bake in a slow oven.

Honey Shortbread

$\frac{1}{2}$ lb all-purpose flour, $\frac{1}{4}$ lb butter, $\frac{1}{2}$ lb honey. Work well together to make a stiff dough. Roll out well, cut into shapes and bake in a slow oven.

Honey Plum Cake

$1\frac{1}{2}$ cupfuls honey, $\frac{2}{3}$ cupful butter, $\frac{1}{2}$ cupful milk, 3 beaten eggs, 3 cupfuls self-rising flour, 2 cupfuls raisins, 1 tablespoon each cloves and cinnamon. Mix into a loaf and bake in a slow oven.

Honey Seed Cake

Beat up $\frac{1}{2}$ cupful butter till light. Add gradually $1\frac{1}{2}$ cupfuls honey, 2 lightly beaten eggs, 1 teaspoonful caraway seeds and 2 cupfuls self-rising flour. Mix well and bake in a moderate oven.

Honey Biscuits

1 lb honey, 2 oz sugar, 2 oz butter, pinch of ginger and a little grated nutmeg. Mix well together and work in enough flour to make a stiff paste. Roll out thinly, cut into shapes and bake on a buttered pan in a quick oven.

Cream Biscuits

$1\frac{1}{4}$ cups honey, $2\frac{1}{2}$ cups sour cream, $1\frac{1}{4}$ teaspoons bicarbonate of soda, any flavouring essence, flour. Beat the honey and cream, add soda and essence and work in enough flour to make a stiff dough. Roll out, cut into shapes, and bake in a moderate oven.

Honey Ginger Snaps

Boil together 1 lb honey, 1 oz butter, $\frac{1}{2}$ oz ground ginger. When nearly cold, stir in enough flour to make a dough, roll out thin and bake in hot oven on a greased plate.

HOME MADE SWEETS

Toffee

Put 4 lb sugar into a pan with 4 cups water. Heat over a clear fire, stirring until it boils, then cover and boil ten minutes. At 310 °F. pour in 1 lb melted butter. Let this boil well in, then add $1\frac{1}{2}$ lb honey, the juice of a lemon and a teaspoonful lemon essence. Boil another minute and pour into greased pans.

Butterscotch

Boil 2 cupfuls honey until a drop sets hard when dropped into cold water. Stir in $\frac{1}{2}$ cupful melted butter, 1 teaspoonful salt and a few drops essence of lemon, almond or vanilla. Pour onto a cool greased plate, cut into squares and wrap in waxed paper when cold.

Honey Fudge

1 cupful each superfine sugar and granulated brown sugar to be boiled in a cupful of milk until it forms a soft ball when dropped into water. Add a cupful of honey and boil again to the same state. Now add $\frac{1}{2}$ oz butter and a

tablespoonful of vinegar, mix well and pour into greased pans. Any flavouring or colouring essence may be added for variety.

Honey Candy

Boil 1 lb honey with $\frac{1}{4}$ lb refined sugar and 2 oz butter until it hardens when dropped into cold water. Turn out into shallow plates and when cool enough to handle, twist or work into desired shapes.

Honey Caramels

1 lb honey, 1 lb sugar, $4\frac{1}{2}$ tablespoons new milk. Boil until it forms a soft ball. Flavour with essence and pour into a greased dish. Before it cools, mark into squares. Wrap in waxed paper when cold.

Honey Ice

2 pints cream, $\frac{3}{4}$ cupful honey, 1 cupful milk, 1 teaspoonful flavouring essence. Warm the milk, add honey and stir till melted. Mix with the cream, flavour and freeze.

Salad Dressing

One part each lemon juice and clear honey, two parts olive oil. Beat well together and add the stiffly whipped white of an egg and a pinch of salt.

HONEY BEVERAGES

Mead or Honey Wine

This is one of the oldest known liquors. When well made it is considered superior to any wine and is beyond doubt a wholesome and fortifying drink.

There are numerous recipes, some of which contain herbs, spices, fruit, etc., but the simple fermented liquor, which should be clear amber, sparkling and not too sweet, is equal to anything more elaborate. It has been customary in the past to use the washings of honeycomb as a basis and to add honey to the liquid until it will float a new laid egg.

If using fresh honey, 4 lb should be added to each 10 pints of water. Some think it an improvement to add a little lemon peel. The mixture is boiled an hour, skimmed and put into a tub or other suitable vessel. Yeast is added either by mixing it first in tepid water, or floating it on the liquor on a piece of toast. When fermentation has started, strain the liquid off into a clean vessel, keeping this lightly covered till it has stopped working.

Small quantities may be made without using a cask, by starting the fermentation in an enamelled bowl and straining off into pint bottles. These should be covered with a piece of cellophane, held on with rubber bands, to allow the gas to escape. Fill up the bottles as needed from time to time. When fermentation stops, decant the liquor into clean bottles and cork securely. After bottling, the mead should stand a year before it is fit for use. Success depends largely on temperature, which should be round about 70 °F. This will be fairly used in a warm kitchen in early autumn.

Honey and Oatmeal

This non-alcoholic drink is made by putting two tablespoonfuls of oatmeal into a quart jug and nearly filling with fresh boiling water. Cover over and stand 24 hours. In another jug dissolve 3 tablespoonfuls of honey in a little boiling water with the juice of two lemons. Strain the oatmeal water into this and it is ready for use. It may be made with pearl barley instead of oatmeal and lime juice instead of lemons.

Honey Vinegar

This is made in various ways, but the basis is about one part of honey by weight to 6 parts of water—1 lb to 6 pints. Some people make it without boiling or adding yeast, keep it in a temperature of about 80 °F. for six weeks, and clear it by adding $\frac{1}{4}$ oz of melted isinglass. If made in summer, it can stand in sunshine all day and be brought in at night.

Others boil the liquor first, ferment it with yeast and add vinegar plant after the fermentation has slowed down. It will not be ready for bottling for some months and when ready, should have a mildly acid taste, with a definite flavour of honey.

HONEY AS MEDICINE

From time immemorial, the virtues of honey as a remedy have been recognized by country folk in all lands and even in these days of patent medicines many people think of honey as the best cure for a sore throat. That it has therapeutic value is indisputable. Its chief value lies in its high nutritive property. All sugars are now established as the most superior food for imparting energy to the body rapidly and a good ration of sugar in some form is indispensable to those who are undergoing severe trials of endurance — long distance runners and swimmers, Arctic explorers and the like. In this regard, honey is superior to cane sugar because it passes directly into the system without the aid of the digestive juices. Hence it will nourish the body under conditions in which even milk fails. It does not, like cane sugar, give rise to gases in a weak stomach, so that not only in an atonic state of that organ, but even when there is ulceration, honey can nourish and stimulate without any unpleasant reaction.

Minute quantities of valuable minerals essential to the proper functioning of secretions—such as salts of iron, phosphorus, lime and sulphur—are contained in honey, and even when the body will not tolerate artificial compounds, those naturally contained in honey are readily absorbed. This easy assimilation, as well as its palatableness, make honey the very best medium for introducing drugs of many kinds, which would be rejected if offered in any other way.

Honey Tea

A tablespoonful of honey in a breakfastcupful of hot water. This should be sipped slowly on an empty stomach. In severe digestive disorders, it will be beneficial to take this several times a day, fasting.

Honey and Lemon Tea

Add the juice of half a lemon to the above. This is very useful in liver disorder and for complexion blemishes. Taken as hot as possible before getting into bed, it will often ward off a cold.

Honey and Yarrow

To an infusion of yarrow, add a good spoonful of honey and drink hot at bedtime and on rising. This has been widely recommended for influenza and as a nutrient tonic.

Honey and Milk

A cup of warmed milk with a teaspoonful of honey. Good for delicate ill-nourished children, and may even be given to infants. Has been highly recommended for cases of stomach ulcer and anæmia and, taken last thing at night, for insomnia.

Honey and Glycerine

A cupful of hot water, 2 teaspoonfuls honey, 1 teaspoonful glycerine. Excellent for colds and sore throat.

Honey and Elderberry

5 spoonfuls elderberry syrup, 1 spoonful honey and a small level saltspoonful of sal prunella (a preparation of fused nitre). Take a spoonful at intervals for sore throat.

Linseed and Honey

Boil one ounce of linseed in $2\frac{1}{2}$ cups water for half an hour. Strain, add the juice of a lemon and sweeten with honey. Take hot at bedtime.

Cough Candy

Boil horehound leaves in soft water, strain through muslin, add as much honey as desired to the liquor and boil until the candy forms a soft ball when dropped into cold water. Pour into greased tins to set.

OINTMENTS AND COSMETICS

Honey has an unassailable reputation as an emollient, while its water-absorbing power makes it of great value in certain skin troubles. Probably its effectiveness in allaying the pain of bee-stings led to its use in other forms of inflammation. The simplest application is to spread clear honey on a cloth and wrap this over the affected area. It is claimed that nothing relieves the pain of erysipelas so readily.

Honey and Glycerine

In equal parts is good for bruises, chafing and chaps on face or hands.

Cure for Chilblains

One tablespoonful of honey is mixed with an equal quantity of glycerine, the white of an egg and enough flour to make a fine paste. A teaspoonful of rose water is helpful.

Wash the affected parts well with a pure soap and warm water, dry thoroughly and spread the paste over. Wrap up with a cotton cloth, as the mixture is very sticky.

This recipe comes from South Africa and it is said that it always relieves and generally cures bad chilblains in a single application.

BEESWAX

Beeswax is a physiological product of the honeybee produced by glands under the abdomen. When gorged with honey and without comb to receive it, the workers cluster closely together, raising the temperature to the region of 90 °F. Within twenty-four hours wax scales begin to form. They are almost transparent and roughly pentagonal in shape. They can often be found on the floor of a hive where they have been dropped during the work of comb building and look like small flakes of mica. The workers pick them from the pockets with the spines on the second pair of legs and transfer them to the jaws, where they are masticated with saliva which makes them plastic. They are then stuck to the roof of the chamber or on the edge of a cell already started and so built up into irregular masses, which in turn are scooped out and thinned down until the familiar cells begin to take shape.

HOW CELLS ARE BUILT

The remarkable precision and regularity with which comb cells are built has been the wonder of mankind for ages, owing to economy in material and perfect adaptation to purpose. The hexagonal shape ensures the least possible waste of space and the construction of cells back to back allows one floor to serve as the base of two cells.

It is now generally considered that this formation is the inevitable result of the method of working. Packed close together, each bee makes a cell the size of its own body by removing wax from the inside until it can reach the bottom. Though this seems a very reasonable explanation, it is not entirely satisfactory, for it does not explain the formation of two different sizes of cells—the workers $\frac{1}{5}$ in and the drone cells $\frac{1}{4}$ in in diameter—though it seems fairly certain that the latter are built when there is a heavy flow of honey, for they will hold a larger quantity with the same amount of wax.

When built during a good flow, the comb is usually white, but the shade depends somewhat on the colour of the nectar and pollen from which it is derived. If sainfoin is the predominant blossom, the comb is deep yellow, while that from heather is snowy white. In light honey-flows, new comb is often darker, because some wax from old comb is mixed with it. As the work proceeds the comb is coloured more or less by the addition of varnish, made from the resinous propolis which the bees collect.

WAX AND HONEY PRODUCTION

There has been much discussion about wax production in relation to honey consumption. At one time figures as high as 30 lb of honey to 1 lb of wax were put forward, but about 8·5 lb is now considered a more accurate estimate. It no doubt varies a great deal according to the weather, strength of the colony and other factors. At any rate, the greater production of surplus honey when bees are provided with ready-built comb shows that wax is costly to produce, so that it is always best to save the bees as much comb building as possible. Even if wax secretion is, as some think, quite involuntary, it is obviously encouraged by retention of food in the stomach and so long as cells are ready to receive honeybees do not retain it long.

MAINTAINING A SUPPLY OF COMB

The aim of the bee-keeper should be to have hives always furnished with the best comb, to retain it as long as possible, consistent with efficiency, and to prevent the formation of 'burr' and 'brace' comb by correct spacing. In spite of all care, there will always be a percentage of wax wasted in various ways. In course of time, combs become inferior for various reasons like change to drone comb, injury by wax moth, or clogging with old dried-up pollen. Brace and burr comb is built outside frames and sections, or in temporary receptacles like skeps and boxes in which bees have been hived during a strong honey-flow. The largest production of wax is from the cappings removed before honey is extracted, and this also provides the best quality.

The amount of wax collected in these ways varies a good deal. If the bee-keeper aims mainly at section production, his wax harvest will be comparatively small, but if he works for extracted honey, the cappings will amount to a considerable quantity. In my own apiary, I collect about 2 lb of wax for every 100 lb of honey. At current prices this means that for every $50 worth of honey there will only be about $1 worth of wax and the more efficient the production of honey, the less important the wax becomes, so that if its collection and preparation involves much trouble it is not worth while and it would be better to destroy it forthwith. By the adoption of simple methods of collection and rendering the trouble is reduced to a minimum and some addition to the income is made. There must be no compromise: either the waste should be destroyed at once by burning or burial, or it should be properly and completely rendered down, because scraps of comb left about can encourage wax moth, one of the worst bee-keeping pests.

Anyway, it is impossible to avoid the work of removing burr comb from hives, frames and sections, or to extract honey without uncapping and it is no trouble to store the waste in an airproof receptacle. 8-lb cookie jars are as useful as anything for the small bee-keeper if he uses a solar extractor, for as soon as the jar is full it can be emptied into the extractor. Care should be taken to keep out as much propolis as possible, for the more this is mixed with wax, the darker will be the colour. In cleaning sections, for instance, burr comb should be cut off and kept as should the scrapings of propolis. Cappings should always be kept separate.

RENDERING WAX—THE SOLAR EXTRACTOR

There are several ways of rendering wax, but the best appliance is undoubtedly the solar extractor, for this not only works without attention or fuel but disposes of the wax waste at once and thus avoids the risk of wax moth attacking the store. I have seen complaints in the bee papers that we do not get enough sun in Britain to operate the extraction properly, but I do not remember a time during the twenty years which I have used one that I did not get all my culled combs and other waste melted down in it.

The appliance can be purchased, but it is easily made at home. The important thing to remember is that its effectiveness depends on the proportion of cubic content to the heating surface.

Get a sound box 24 in long, 15 wide and 10 deep. Line it with something to retain heat. Next fit inside the upper part, a tray 6 in deep, the full width of the box and three-quarters its length. The tray can be made of zinc or

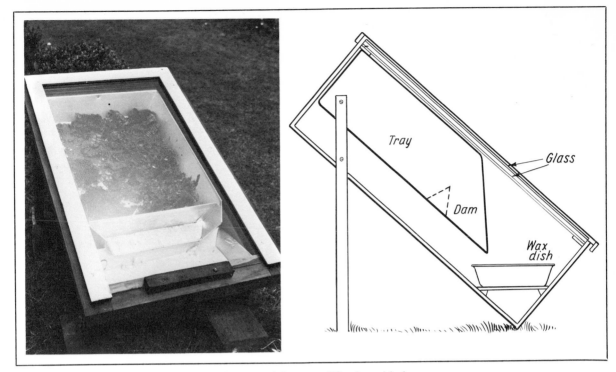

Above left: solar wax extractor
Above: diagram of home-made solar wax extractor

wood. I have found stout beaverboard very satisfactory. The free side is not closed in, but shaped to form a chute with its mouth about 6 in across, and a piece of excluder zinc is cut to fit inside this end to prevent rubbish falling off. It must be loose, so that it can be taken out for cleaning.

The box must be tilted to an angle of 45° and a pair of legs fitted to keep it in that position. Under the mouth of the tray a dish of some kind is placed to receive the wax, a piece of wood being fixed beneath it so that it stands level.

The cover is made of a wooden frame 2 in wide and ½ in thick, fitted comfortably outside the box. Round the inside of this frame, nail 1 in lath, exactly in the centre. Two pieces of glass are cut to fit on this rebate: one inside, kept in place by narrow strips of wood tacked to the sides; and one outside, puttied in to keep out rain. The use of two glasses an inch apart is the secret of this extractor, for a single glass will allow heat to escape quickly if the sun is obscured. To lift the cover easily on and off, fasten a C-shaped handle at the top end. If well made and painted the apparatus will last a lifetime.

This appliance stands in the open all summer, facing south. The waste wax is put into the tray and without further attention runs into the dish. When sufficient wax has accumulated, it should be poured out while hot into some sort of mould. Nothing is better than an earthenware jam pot. Before pouring the wax into it, first fill it with water, empty this out and fill with wax before the inside dries. When set, the wax will turn out in a solid cake. From time to time, the rubbish in the tray should be cleared out while still warm. A sheet of paper laid under the chute will receive it. This 'slum gum' should not be thrown away, but allowed to harden. If broken in pieces about the size of a golf ball, it will burn long and fiercely, and forms an excellent pleasant-smelling fire-lighter.

Steam wax extractor

OTHER METHODS

Wax may also be melted in water, the commonest method being to pack the comb into a muslin bag, tie it securely and put it in a pan or copper of water, with a weight on it to keep it down. The water is boiled to extract the wax, which floats to the surface and can be removed in a cake when cool. It is essential to use rain water, because any trace of lime will spoil the texture of the wax, but a little vinegar added to hard water will counteract this.

However, this method tends to discolour the wax owing to the resins and other substances boiled with it and it is better to use a steam extractor, several patterns of which are on the market. In this type (left) the wax does not come into contact with the water, but runs off as it is melted by the steam.

In the style illustrated below left, the melted wax is forced through the strainer top by pressure from water poured in the tube.

DEALING WITH CAPPINGS

Wax extractor: water pressure forces the melted wax through the strainer

I mentioned that cappings produce the best wax. The manner in which they are dealt with depends on the size of the apiary. Large bee-keepers usually have an uncapping appliance which automatically melts and separates the wax. The cappings fall onto a sloping tray which is heated beneath by hot water, steam or electricity, so that the honey runs off and the melted wax follows. In the more costly types, the wax runs into a separate receptacle, but in the cheaper kinds it forms a cake on top of the honey.

Those who do not have much to spend on appliances have two alternative methods of dealing with cappings. After they have been well drained in the straining tank, they may be put in a pan of tepid water, soaked for twenty-four hours and drained off again. The liquor can then be used to make mead or honey vinegar and the wax dried off and put into the solar extractor.

Another method is to use the cappings' cleaner devised by Mr Bowen. This consists of a box with metal bottom, supporting a tray of wire netting, $\frac{1}{2}$ in mesh, into which the wet cappings are placed. The box is put on the hive like a super and the bees enter from openings beneath. In a few days they clean out all the honey, leaving the wax as a dry powder. The only objection to this plan is that if there should be brood disease in any hive, there is grave risk of spreading it to the one which receives the cappings.

Cappings wax is much lighter in colour than that from mixed waste, and all gum should be cleared from the solar extractor before it is put into the tray; it is the best material from which to prepare wax for exhibition or pharmaceutical purposes.

MAKING FOUNDATION

Having procured the cakes of wax, either by solar or water extraction, the simplest plan is to send them to the appliance dealer to be made into foundation, but apart from any monetary profit, it is interesting to work up beeswax into various forms. It is possible to make foundation at home, but it is troublesome and the product is never so good as that from the factory. It is made by first dipping a wet board into a vat of melted wax, thus coating it on both sides with a thickness depending on the number of times the board is dipped.

178

When cold, the wax can be stripped off in a sheet. To make the impression of cell bases the wax sheet is passed through rollers operating like a mangle, bearing the matrix. The apparatus is costly and very few, even of the larger bee farmers, consider the outlay justified. However, metal moulds for producing single sheets of foundation may now be obtained from the appliance manufacturers. Most of these are made abroad and are costly for the individual to purchase but are quite within the capabilities of many bee-keeping associations.

Cakes produced by the water method have a mass of wet grains on the under-side. These consist of pollen, which swells in the water and is slightly heavier than wax. In solar extraction some pollen is mixed with the wax, while at the bottom of the dish there is often a small amount of propolis which has become liquid and run off. Before moulding, the wax must be freed from this gum and pollen. The cake is broken up into a pan of clean rain water and boiled for a few minutes. On cooling, wet pollen will be found on the underside and should be scraped off. The gum dissolves into the water.

This produces a cake of lighter colour, but it may be bleached still more by putting it in the solar extractor again. These processes may be repeated, but the improvement is less each time and it is not worth while to do it more than twice.

MOULDING WAX

For retail purposes, wax should be moulded into cakes of 1, 2, 4 or 8 oz. Suitable moulds of metal or wood can be had from the dealers, but small glass pots, cups, etc., serve well. An egg cup, for instance, makes an ounce cake appropriately shaped like a skep.

To melt the wax, the only safe way is to use a water bath. Special double vessels are sold for the purpose, but it suffices to use a deep can or jar. Break the cakes into small pieces, put them in the can, stand this in a pan of hot water and boil. As the wax melts, more can be added until the jar is full. Meanwhile the moulds should be prepared. They must be quite clean but need not be greased, though sticking is less likely if a little olive oil is rubbed over the inside. It is important to have the moulds warm or the cakes may crack on cooling. There is no better plan than to stand them in a pan of hot water, so that cooling is gradual.

When ready, lift the pot of wax from the water and wipe the outside to prevent drops of water falling into the mould and causing pits in the cake. Pour the wax slowly and evenly to avoid air bubbles, but if any appear on the surface they should be broken while the wax is still liquid. When a crust has formed over the cake, more warm water should be poured into the pan, until it flows over the mould and the wax is quite covered. When cold the cake will float out of the mould.

EXHIBITION WAX

The quality of wax at honey shows has risen greatly during the last few years and the standard must be very good to win prizes. Purity, colour, texture, aroma and perfect moulding are the requirements and great care and attention are needed to form a perfect cake of wax.

Those who intend to exhibit should set aside all pieces of virgin wax, as well as cappings from the whitest comb. Heather cappings are the best and, at large shows, wax from these has a class to itself. Apart from using special care to eliminate propolis, which gives the wax a resinous odour, the procedure is as already described, the wax being rendered in the solar extractor, boiled in rain water to remove pollen, and, if not light enough, passed through the solar extractor again to bleach it. It must be remembered that too much bleaching and refining will spoil the aroma. When enough refined wax has accumulated, pick out rather more than a pound of the best, for a show cake must usually be about 1 lb in weight.

Melt the selected wax in an earthenware jar as usual. Meanwhile, get a piece of old well-shrunk flannel, large enough to cover another jar and leave a depression in the middle into which the wax can be poured. It can be tied to the jar or held in place with a rubber band. Stand this jar in the oven and make it as hot as the hand will bear. After wiping moisture from the jar of melted wax, pour it into the flannel. It will run through easily, leaving a surprising amount of dirt on the flannel. Withough delay remove the strainer and pour the wax into the mould, which should be standing ready in hot water. Needless to say, the mould must be as perfect and well polished as possible, and the wax must be poured in smoothly to ensure a perfect cake.

Exhibition wax must be in a plain cake, but it may be round, square or oblong. Suitable moulds can generally be found in the pantry among household soup plates and dishes.

Those with artistic leanings will often take delight in moulding wax into fancy shapes; candles are now encouraged in competitive wax classes, and they form an attractive addition to composite displays of bee produce. Cakes moulded into shapes of animals and flowers look very attractive, and there are few households which have not some pieces of moulded earthenware or glass which may be used to form such designs. Those skilled at wood carving can also make moulds from fine-grained hard wood, and from one such mould any number of casts can be taken. Thin sheets of wax with a moulded pattern can also be produced by dipping the mould into melted wax one or more times till the required thickness is obtained.

The following are useful recipes for household articles:

Beeswax and Turpentine Polish

Melt 1 lb of beeswax and stir in $2\frac{1}{2}$ pints turpentine while the wax is cooling. The exact quantities do not matter, as more turpentine can be added if the mixture is too thick.

Furniture or Shoe Cream

8 oz beeswax, 1 oz white wax, 1 oz Castile soap. Cut all in pieces and boil twenty minutes in a quart of rain water. When nearly cold, add $2\frac{1}{2}$ pints turpentine and shake till a good cream is formed.

Black Wax

2 oz beeswax, $\frac{1}{2}$ oz burgundy pitch. Melt together and add $1\frac{1}{2}$ oz fine ivory black.

Waterproofing for Shoes

4 oz beeswax, 4 oz resin, 1 pint linseed oil, $\frac{1}{4}$ pint turpentine. Melt wax and resin together over gentle heat and stir in the oil. Remove from the fire and add the turpentine. When required for use, melt and rub well into the leather.

Harness Dressing

2 oz mutton suet, 6 oz beeswax, 6 oz powdered red sugar candy, 2 oz soft soap, 1 oz indigo or lamp black. Dissolve soap in $\frac{1}{4}$ pint water, add the other ingredients, melt, and mix together. Then add $\frac{1}{4}$ pint turpentine. Apply with a sponge and polish with a brush.

Fruit Bottle Covers

2 oz beeswax, 4 oz resin, $\frac{1}{2}$ oz vaseline. Melt all together in a tin and brush evenly over pieces of linen or calico. When required for use, apply to the hot jar and press down firmly.

Note that in the recipes that include turpentine, pure turpentine is called for, rather than the white spirit or 'turpentine substitute' often used for thinning household paints.

17
Honey
Shows

It is natural that those engaged in the arts and crafts should take pride in their work and find it a joy to exhibit their best achievements to their fellows. In modern times this trait is fostered in public exhibitions of all kinds, and there is an annual round of agricultural and horticultural exhibitions all over the country. These shows serve a valuable purpose by setting a high standard of production as the aim for all. The American Honey Show, held in conjunction with the American Beekeeping Federation Convention, is the most important of these exhibitions, and takes place each year in late January. Silver trophies and Blue Ribbons are awarded to all the different classifications.

EXHIBITION QUALITIES

Generally speaking, the production of honey, etc., for exhibition does not differ from the best methods of normal production, but for show purposes, the finest samples are naturally chosen and special care taken to present them in the best possible manner.

Extracted honey usually comprises the larger number of exhibits, and at all but the smallest shows is divided into several classes. Liquid honey is classified as light, medium and dark, and the first difficulty of the exhibitor is to decide into which of these his sample comes. It is done by reference to a standard colour glass of a specific yellow tint: any sample which is lighter than this glass falls into the light class; any honey which is darker than the tint of two such glasses placed behind each other, goes into the dark class; all those between are in the medium class. This is the grading system which prevails in Britain. US classifications are given under Liquid Honey below.

Except in certain districts, such as the downlands on chalk soils where honey is consistently light year by year, the colour of honey varies infinitely from place to place, season to season and even month to month. Light honey, generally esteemed the best, comes mainly from clover, and other things being equal, some judges prefer even lighter specimens. Others prefer the bright yellow of sainfoin.

Those whose crop comes from varied sources must make a selection of the combs before extracting. It is easy to see whether the honey is light or dark by holding the comb to the light. The light ones are then extracted first and the honey kept in a separate can.

All combs chosen for show honey should be fully sealed, otherwise the next most important factor, density—or as it is more correctly called, viscosity— will fall below standard. Any honey which is thin and runs freely after uncapping should be rejected for show purposes. It should stick to the knife, even though that tool has been properly warmed.

The selected can of honey should be warmed by standing it in a vessel of hot water and then the honey run through a bag made of old flannel, or alternatively through a clean nylon stocking into a clean vessel, in which it should stand a few days, so that any scum which rises to the surface can be skimmed off before the honey is bottled.

CONTAINERS

Highly plated lids are barred from some shows, but to ensure the best appearance the clearest jars, free from flaws and scratches, should be chosen. It is a minor point, but in a closely contested class, it may turn the scale.

Once the jars have been selected and well polished inside and out, the strained skimmed honey can be poured in right to the top and a temporary covering put on. The jars are then allowed to stand for a week in a warm place: a sunny window or greenhouse is very suitable. The honey at the top should then be taken off with a spoon down to the neck ring, and any trace of scum round the edges carefully wiped off before new clean caps are put on. Before sending them away, the jars must be well polished and the labels provided by the show secretary affixed. No private label, or anything of a distinctive character, is allowed on a jar.

JUDGING

There are two schools of opinion among judges of honey. One claims that the only fair way to decide is by a system of points allowed for each factor, while others maintain that such a system is impossible and that judging the best of a batch is just a matter of skill and experience. Both schools agree that flavour is the most important factor, yet many judges have been known to reject the best-flavoured exhibit, because it contained some smaller fault—fragments of bees, or scum.

LIQUID HONEY

Liquid honey is classified into several categories: water white, extra white, white, extra light, light amber, amber and dark. Each classification is judged by the purity of its appearance, its uniformity and the absence of granulation.

CREAMED (GRANULATED) HONEY

Only light or medium honey produces a satisfactory show honey in this class, as dark honey crystallizes to a brown muddy colour. It is rarely possible to produce a good sample without special treatment, for even though it granulates finely, there is nearly always a certain amount of 'frosting' which spoils the appearance. The granulated honey should first be prepared in the same way as for the liquid classes and then heated to a point between 140 and 150 °F. While still warm, it is stirred steadily until it begins to thicken, when it should be poured *slowly* into the jars to avoid air bubbles. The jars should then stand in a warm place for a few days. Any scum should then be removed, after which the jars should be exposed to light until the honey sets white and firm.

COMB HONEY

Comb honey is exhibited in three forms: sections, cut comb and combs for extracting.

Sections

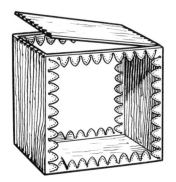

Show case for sections

The sections require careful preparation from the start. A full sheet of foundation is fitted with great care, so that it does not buckle and each section should be quite square in the rack with dividers placed evenly between the rows.

Good sections can rarely be produced except by a strong stock in a good honey-flow, so that all are quickly filled with one variety of honey only and sealed up rapidly. In hot weather a good stock will fill and cap over a whole section rack in a week or ten days. As soon as they are sealed over, the sections should be removed, for the bees will continue to add wax to the cappings and their journeys to and fro tend to soil the surface.

Special care must be taken in cleaning sections for show, and to facilitate the removal of propolis, it is a good plan to coat the sections with paraffin wax before making up. At one time sections used to be glazed and furnished with fancy edging for show, but as judges insist on being able to taste the honey, which they should do by opening one cell only and inserting a glass tube, it is now usual to put them in special cases sold by the appliance dealers.

Show sections must be between 16 and 17 oz in weight, filled with one kind of honey only, all cells sealed over with even capping, and translucent by transmitted light. There must be no popholes or pollen in them. Generally speaking, the whiter the comb, the higher it is classed, but the bright yellow of sainfoin is just as highly marked, other things being equal.

Cut Comb and Chunk honey

Both are popular entries at honey shows.

Extracting Combs

Show case for shallow combs

These may be of shallow or standard size—sometimes there is a separate class for each—and should be of virgin comb. They must be well filled, but not excessively bulgy, and the comb surface should be flat and free from stains and markings. Cells with pollen in them are the chief thing to avoid. Special glazed cases are sold by the appliance dealers to hold show combs, but they can also be made at home by anyone handy with tools.

BEESWAX

This must be in a plain cake, generally 1 lb in weight. Exhibitors usually have a glass case made to fit it, nicely lined to show off the colour, which may range from pale primrose yellow to deep gold.

18
Directory of
Apiculture

The following lists should be helpful to the reader, but they are necessarily selective and he is advised to subscribe to one or more of the trade publications, join one or more of the organizations and become acquainted with the types of merchandise and services offered.

The major suppliers of hives, supers, frames, extractors, smokers, honey pumps, comb-honey sections and a complete line of bee-keeping equipment are as follows:

A. I. Root Co., Medina, OH 44256 (with branches in other States)
Dadant & Sons, Inc., Hamilton, IL 62341 (with branches in other States)
Leahy Mfg. Co., 406 W. 22nd Street, Higginsville, MO
Walter T. Kelley Co., Clarkson, KY 42726
August Lotz Co., Boyd, WI 54726

Queen and package-bee producers include:
Dadant & Sons, Inc., Hamilton, IL 62341 (the originators of Starline and
    Midnite Hybrid Queens. Also available from numerous queen breeders)
York Bee Co., Box 307, Jesup, GA 31545
Weaver Apiaries, Rt. I Box 111, Navasota, TX 77868
Howard Weaver & Sons, Rt. I, Box 16-A, Navasota, TX 77868
C. G. Wenner & Son, Rt. I, Box 283, Glenn, CA 95943
E. J. Bordelon Apiaries, Box 33, Moreauville, LA 71355
Bordelon Apiaries, Rt. I, Box 151, Kasson, MN 55944
Williams Apiaries, Rt. I, Box 6-B, Macon, MS 39341
Bee Island Apiaries, Harrells, NC 28444
Shumans Bee Co., 407 Jefferson Street, Hazelhurst, GA 31539
Rossman Apiaries, Inc., Box 905, Moultrie, GA 31768
Calvert Apiaries, Calvert, AL 36513
The Stover Apiaries, Inc., Mayhew, MS 39753
Wilbanks Apiaries, Box 12, Claxton, GA 30417
Clear Run Apiaries, Box 27, Harrells, NC 28444
Walter T. Kelley Co., Clarkson, KY 42726
Keyline, Inc., Box 218, Theodore, AL 36582
Gulf Coast Bee Co., Schriever, LA 70395
Strachan Apiaries, Inc., 2522 Tierra Buena Road, Yuba City, CA 95991
Kane Apiaries, Rt. I, Box 192, Hallettsville, TX 77964
Geo. E. Curtis & Sons, Inc., Box 507, La Belle, FL 33935

Producers national marketing cooperative organizations

Sioux Honey Association, Box 1107, Sioux City, IA 51102 (with branch offices in a number of honey-producing areas)

Midwest Honey Producers Association, Howard Schmidt, Winner, SD 57580

Honey packers and/or dealers

Fischer Honey Co., 2008 Main Street, North Little Rock, AR 72114

Howard Foster, Box 239, Colusa, CA 95932

Superior Honey Co., 10920 S. Garfield, South Gate, CA 90280

Tropical Blossom Honey Co., Box 8, Edgewater, FL 32032

Wayne Budge, 220 West 500 North, Malad, ID 83252

W. F. Straub & Co., 5520 Northwest Highway, Chicago, IL 60630

Robert J. Steele, 3201 Bushnell Ave., Sioux City, IA 51106

Hubbard Apiaries, Onsted, MI 49265

R. B. Willson, Inc., 250 Park Ave., New York, NY 10017

Wayne Stoller, Box 97, Latty, OH 45855

Burleson's, Inc., Box 578, Waxahachie, TX 75165

Weaver Apiaries, Rt. I, Box 111, Navasota, TX 77868

Vanderford-Blackwell Apiaries, 715 E. 2nd Ave., Ellensburg, WA 98926

Bee-keeper organizations

American Bee Breeders Association, L. C. Harbin (Sec.-Treas.), Box 218, Theodore, AL 36582

American Beekeeping Federation, Robert Banker (Sec.-Treas.), Rt. I, Box 68, Cannon Falls, MN 55009

American Honey Institute, Leslie Little (Sec.), 831 Union Street, Shelbyville, TN 37160 (a division of the Honey Industry Council of America)

California Honey Advisory Board, Manager-Home Economist, Mona Schafer, Box 32, Whittier, CA 90608

Canadian Department of Agriculture, Ottawa Research Station, Central Experimental Farm, Ottawa KIA-OC6

Honey Industry Council of America, John Root of A. I. Root Co., Medina, OH 44256

Ladies Auxiliary of ABF, Lois Blackwell (President), 715 E. 2nd Ave., Ellensburg, WA 98926

Minnesota Beekeepers Association, Box 6201, Rochester, MN 55901

Bee-keeper publications

The American Bee Journal, Hamilton, IL 62341

The American Beekeeping Federation News Letter, Rt. I, Box 68, Cannon Falls, MN 55009

Gleanings in Bee Culture, Medina, OH 44256

The Speedy Bee, Rt. I, Box G-27, Jesup, GA 31545

Details of your State association can be obtained from your local apiary inspector. In addition, two of the leading magazines mentioned above, *Gleanings in Bee Culture* and the *American Bee Journal*, also publish detailed lists of names and addresses of State associations.

19
Literature of Bee-keeping

Few subjects of human interest have such a rich literature as that of bee-keeping. Quite apart from its economic importance, which is much greater than is commonly supposed, the honeybee fascinates her admirers by her remarkable organization, the beauty of her structures and the devotion with which she labours, indoors and out, to ensure the prosperity of the race. Nor is this interest lessened, but rather increased, by the wholesome respect which people have for her defensive armament. 'They came about me like bees', says the Psalmist, intimating the most intolerable persecution he could contemplate. 'My son, eat thou honey, because it is good', says Solomon. 'Butter and honey shall he eat, that he may know to refuse the evil and choose the good'—Isaiah.

Many writers of the Classical period refer to bees: Aristotle (342 BC), Cato (200 BC), Varro (100 BC), and particularly Virgil (50 BC), whose fourth Georgic shows plainly that he was a practical and enthusiastic bee-keeper. Columella (AD 60) and Palladius (AD 350) also wrote about bees in a manner which shows that they had practical knowledge of the subject.

The earliest book in English to mention bees is *The Arte of Gardening* by Thomas Hill (1574) which contains a treatise on bees, but it is not very original, since it consists mostly of quotations from the classical writers. In 1609, however, Charles Butler wrote a really original book called *The Feminine Monarchie* in which he corrected the ancient authors who always alluded to the 'King' bee, by showing that the ruler was 'a queen and mother of all the others'. Two other notable seventeenth-century books were Jacob Nickel's *Die recht bienenkunst* (1614) and John Gedde's *New Discovery of Bee Houses* (1675).

In the eighteenth century quite a lot of books on bees were written and the following have definitely influenced the course of the craft.

1712	Warder, Joseph	*The True Amazons*
1734	Réaumur, R. A. F. de	*Mémoires pour servir à l'histoire des insectes*
1744	Hornbostel, H. C.	*Wie das wachs von den bienen kömt*
1744	Thorley, John	*Female Monarchy*
1758	Swammerdam, Jan	*Book of Nature*
1768	Wildman, T.	*Management of Bees*
1771	Schirach, A. G.	*Histoire Nat. de la Reine des Abeilles*
1771	Janscha, Anton	*Abhandlung von Schwärmen der Bienen*
1780	Keys, John	*Practical Bee-master*
1781	Dyer, W.	*The Apiary Laid Open*
1789	Bonner, James	*Bee-master's Companion*
1790	Rocca, l'abbé della	*Traité complet sur les abeilles*

At the very end of the century the most famous bee book of all appeared. This was the *Nouvelles observations sur les abeilles* of the blind Swiss, François Huber. That book began an entirely new epoch in bee-keeping, for it was his discovery that it was possible to make bees build comb in separable frames, that laid the foundation of the modern system. That was by no means the only contribution Huber made to progress in apiculture. He revealed, for the first time, almost the entire story of the queen's life and work, finally settled the vexed question of the origin of wax, ascertained just how the bees keep the hive ventilated and proved that pollen is required mainly for breeding purposes. A very good revised translation of Huber's great work was published in 1926—*New Observations on Bees* by C. P. Dadant.

The nineteenth century produced a shoal of books about bees, both in Britain and America. The most noteworthy was that of Langstroth, 'Father of American bee-keeping', *The Hive and the Honeybee* (1853), in which he made clear the secret of the 'bee space'. This book went through many editions, but has since been re-written by a team of specialists.

Dzierzon's *Rationelle Bienenzucht* (1861) was another work of outstanding importance. This was translated into English by C. N. Abbott in 1882. In England the most valuable book of the century was Cheshire's *Bees and Bee-keeping* published in two volumes in 1886, but the work which has, perhaps, been the most helpful, both in Britain and America, was the rival to Langstroth's *The Hive and the Honeybee* written in 1883 by A. I. Root and called *The A.B.C. of Bee Culture*. Other books of the century were:

1815	Huish, R.	*Treatise on Bees*
1827	Howatson, T. M.	*Apiarian's Manual*
1827	Bevan, Edward	*The Honey Bee*
1832	Nutt, Thomas	*Humanity to Honeybees*
1834	Bagster, Samuel	*Management of Bees*
1853	Quinby, M.	*Mysteries of Bee-keeping*
1862	Woodbury, T. W.	*Bees and Bee-keeping*
1865	Neighbour, A.	*The Apiary*
1881	Cowan, T. W.	*British Bee-keeper's Guide Book*
1887	Simmins, S.	*Modern Bee Farm*
1889	Doolittle, G. M.	*Scientific Queen Rearing*
1890	Cowan, T. W.	*The Honey Bee*

Quite early in the present century a book which has earned the widest popularity appeared. It was Count Maurice Maeterlinck's *La Vie des Abeilles*. Translated into English by Alfred Sutro, *The Life of the Bee* has sold more than 100,000 copies.

As in many other spheres, there has been a tendency for writers about bees to specialize in various directions and for the benefit of those who wish to pursue a particular side of the subject, I have divided the twentieth-century books into those classes.

BEE LIFE

1908	Edwardes, Tickner	*Lore of the Honeybee*
1921	Mace, H.	*Book about the Bee*
1933	Rendl, G.	*Way of the Bee*
1939	Francon, Julien	*Mind of the Bees*

1943	Teale, E. W.	*The Golden Throng*
1946	Kelsey, W. E.	*Spell of the Honeybee*
1948	Metcalf, F. H.	*The Bee Community*
1948	Wadey, H. J.	*Behaviour of Bees*
1950	Butler, C. G.	*The Honeybee*
1964	Ribbands, C. R.	*The Behaviour and Social Life of Honeybees*
1968	Frisch, K. von	*Bees: Their Vision, Chemical Senses and Language*
1974	Butler, C. G.	*World of the Honeybee*

PRACTICAL BEE-KEEPING

1903	Miller, C. C.	*Forty Years Among Bees*
1904	Digges, J. G.	*Practical Bee Guide*
1915	Phillips, E. F.	*Bee-keeping*
1924	Sturges, A. M.	*Practical Bee-keeping*
1927	Mace, H.	*Modern Bee-keeping*
1931	Herrod-Hempsall, W.	*Beekeeping New and Old*
1931	Lawson, J. A.	*Honeycraft*
1936	Manley, R. O. B.	*Honey Production in the British Isles*
1938	Dadant, C. P.	*First Lessons in Bee-keeping*
1939	Hooper, M. M.	*Commonsense Bee-keeping*
1943	Wadey, H. J.	*Bee Craftsman*
1945	Hamilton, W.	*Art of Bee-keeping*
1946	Manley, R. O. B.	*Honey Farming*
1947	Whitehead, S. B.	*Honeybees and their Management*
1948	Manley, R. O. B.	*Bee-keeping in Great Britain*
1950	Cumming and Logan	*Beekeeping, Craft and Hobby*
1950	James, E. L. B.	*Beekeeping for Beginners*
1963	Grout, R. A.	*The Hive and the Honeybee*
1975	Wedmore, E. B.	*Manual of Bee-keeping*

SWARMING

| 1934 | Snelgrove, L. E. | *Swarming* |
| 1946 | Bent, E. R. | *Swarm Control Survey* |

ANATOMY AND PHYSIOLOGY

1923	Betts, Annie D.	*Practical Bee Anatomy*
1956	Snodgrass, R. E.	*Anatomy of the Honey Bee*
1962	Dade, H. A.	*Anatomy and Dissection of the Honeybee*

HISTORICAL

1931	Fraser, H. M.	*Bee-keeping in Antiquity*
1937	Ransome, Hilda	*The Sacred Bee*
1938	Pellett, F. C.	*History of American Bee-keeping*
1958	Fraser, H. M.	*A History of Beekeeping in Britain*

DISEASE

| 1911 | Zander, E. | *Die Faulbrutt und ihre Bekämpfung* |

1927	Rennie, John	*Acarine Disease in Hive Bees*
1934	Apis Club	*Bee Diseases*
1963	Bailey, L.	*Infectious Diseases of the Honeybee*
1967	U.S. Dept. Agr.	*Identifying Bee Diseases in the Apiary*

QUEEN AND BEE BREEDING

1903	Alley, H.	*Improved Queen Rearing*
1905	Sladen, F. W. L.	*Queen Rearing in England*
1918	Pellett, F. C.	*Practical Queen Rearing*
1923	Smith, Jay	*Queen Rearing Simplified*
1927	Watson, L. R.	*Controlled Mating of Queen Bees*
1940	Snelgrove, L. E.	*Introduction of Queen Bees*
1946	Snelgrove, L. E.	*Queen Rearing*
1947	Abbott, C. P.	*Queen Breeding for Amateurs*
1950	Laidlaw and Eckert	*Queen Rearing*
1968	Brother Adam	*In Search of the Best Strains of Bees*
1975	Brother Adam	*Bee-keeping at Buckfast Abbey*

BEES AND PLANTS

1930	Pellett, F. C.	*American Honey Plants*
1938	Allen, M. Yates	*European Bee Plants*
1946	Howes, F. N.	*Plants and Bee-keeping*
1947	Harwood, A. F.	*British Bee Plants*
1949	Mace, H.	*Bees, Flowers and Fruit*
1956	Lovell, H. B.	*Honey Plants Manual*
1974	Hodges, D.	*Pollen Loads of the Honeybee*

ALLIED SUBJECTS

1938	Beck, B. F.	*Honey and Health*
1948	Gayre, G. R.	*Wassail in Mazers of Mead*
1949	Gregg, A. L.	*Philosophy of Beekeeping*
1973	Imms, A. D.	*Insect Natural History*
1975	Crane, E. E.	*Honey—a Comprehensive Survey*

Many of the books subsequent to 1840 can be obtained through the usual libraries. The most extensive bee-keeping libraries in North America are:

The Bee Culture Branch of the National Agricultural Library at the US Department of Agriculture, Beltsville, MD

The Phillips Memorial Beekeeping Library at Cornell University, Ithaca, NY

The University of California, Davis, CA

The University of Minnesota, Minneapolis, MN

The C. C. Miller Memorial Beekeeping Library at the University of Wisconsin, Madison, WI

The University of Guelph, Ontario, Canada

On an international scale, the library of the International Bee Research Association at Hill House, Gerrards Cross, Bucks., England, has a wide-ranging collection.

Index